THE
BODY LANGUAGE
AND
EMOTION
OF
CATS

Also by Myrna M. Milani
The Body Language and Emotion of Dogs

MYRNA M. MILANI, D.V.M.

THE

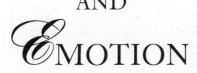
BODY LANGUAGE

AND

EMOTION

OF

CATS

QUILL
WILLIAM MORROW
NEW YORK

Library of Congress Cataloging-in-Publication Data

Milani, Mryna M.
    The body language and emotion of cats.

    Includes index.
    1. Cats—Behavior. 2. Cat owners—Psychology.
3. Human-animal relationships. 4. Emotions. I. Title.
SF446.5.M55 1987 636.8´0887 87-5772
ISBN 0-688-12840-8

Printed in the United States of America

First Quill Edition

1 2 3 4 5 6 7 8 9 10

BOOK DESIGN BY MARIE-HÉLÈNE FREDERICKS

In memory of Eunice Riedel,
who combined her great love of nature,
understanding of the power of the human/animal bond,
and editing skills,
leaving writer and reader with this very special legacy

$\mathscr{A}$CKNOWLEDGMENTS

Michael Snell and Eunice Riedel, whose encouragement, support, and editing make them both worthy of this paraphrase of Lao-tzu

Good editors are best when writers barely know they exist
Not so good when writers always obey and acclaim them
Worst when writers despise them.
Of good editors, when their work is done and their aims fulfilled,
The writer will say, "I did this myself."

This book says what I want it to say because of their help.

Brian Smith, spouse and co-owner of Maggie the cat, who taught me awareness of many other feline dimensions, including their ability to play football in a tiny house, ride motorcycles, enjoy Handel and Mozart, intimidate dogs with the subtlest motions, and unerringly know when humans are busy so that they can streak to the door to demand to be let out or in.

Dick Webber of the Fitzwilliam Depot Store and all the other cat owners who over the years have asked, "Why are cats so different?" and his cat, Albert, and all the other great cats that made us humans think such deep thoughts.

CONTENTS

THE
BODY LANGUAGE
AND
EMOTION
OF
CATS

ℐNTRODUCTION

Comedian George Carlin performs a funny routine called "Dogs and Cats" that brings to mind William Kunstler's observation, "A dog is like a liberal. He wants to please everybody. A cat really doesn't need to know that everybody loves him." When Carlin plays the dog, he perfectly portrays the floppy, sloppy, tongue-lolling pooch who wiggles and wags when the owner comes home, whether after a minute's or a day's absence: "You're home! You're home! I thought you were going to be gone forever and I don't know how to work the can opener, so I rolled this can of food all over the kitchen floor. I love you! I love you!" Then Carlin slips into the role of the cat, blinking haughtily as if trying on a new pair of eyes, and murmuring, "Oh, it's you. Big deal. I killed a mouse for supper. But, hey, if it turns you on, I'll be glad to rub my lean, beautiful body against your legs."

Dog and cat people alike laugh uproariously at the routine, but often for completely different reasons. Dog people relish Carlin's ability to describe the intensity of canine devotion, especially when compared to what they see as the cat's supercilious detachment. Simultaneously, cat lovers delight in Carlin's ability to contrast the typical elegant, independent, and controlled feline behavior with what they perceive as that of a slavish, slobbering dog.

Similarly paradoxical human attitudes toward cats and dogs capture the imagination of two of the most well-known American cartoonists,

Introduction

Charles Schultz, the creator of the "Peanuts" comic strip, and Jim Davis of "Garfield" fame. Schultz presents Snoopy the beagle as the epitome of infinite canine qualities, yet few probably even remember the cat, Feron, who was invariably draped like a rag doll over his mistress's arm. On the other hand, Davis portrays Garfield as the infinitely creative feline and his companion, Hobie, as a mindless canine twit. Even though both artists reflect entirely different views, cat owners have little difficulty accepting both.

As if this weren't confusing enough, we have tongue-in-cheek books describing all sorts of irreverent uses for postmortem felines that convulse some cat lovers with laughter and others with rage. Or consider a revelation made by best-selling author and radio personality Garrison Keillor when he appeared before the National Press Club. It was a foregone conclusion that the subject of "Bertha's Kitty Boutique" was bound to come up. Bertha's, one of the most popular of the imaginary sponsors of Keillor's radio show, "The Prairie Home Companion," offers cat lovers a wide variety of products and services from cat sneakers and sun screen to charm school, guaranteed to convert your everyday cat to a star or your money back.

As I listened to all the Bertha's commercials, like many I envied Keillor's ability to cut through the mystique and get to the heart of what was really going on in the most intimate corners of the feline mind. I marveled at his flawless analysis of many human/feline interactions. At times, his observations were so incredible, it seemed he must surely be part cat himself, a cat lover of extraordinary sensitivity.

Imagine how shocked I was when Keillor revealed rather sheepishly that Bertha's began as a spoof on the antics of cats and cat owners, both of which he finds totally incomprehensible! What he considered the most outlandish parodies, what should have struck us cat lovers as blatant attacks, had exactly the opposite effect. Countless cat lovers recognized bits and snatches of ourselves and our cats in Bertha's ads and laughed uproariously—and heaved sighs of relief, glad to know our relationships with our cats weren't the only bizarre ones.

Introduction

However, having studied feline behavior and human/feline interactions, it shouldn't have surprised me that such incredible insight comes from a person who defines himself as "noncat." *All* aspects of the human/feline relationship are riddled with such paradoxes, some of them even more unusual and most seemingly unresolvable.

In this book we're going to explore how basic feline behavior and body-language expressions make it impossible to evaluate cats in human terms. Regardless of how hard we try, there will always be enigmas and paradoxes that are both unresolvable and uniquely feline. Although we might long to clarify our relationships with our feline friends by amassing support for interpretation A *or* B as the right way for us and our cats to respond, time and time again the cat demands that we accept A *and* B—and even any C, D, and E. Moreover, we will discover that even though we can accept A or B as true for ourselves and our cats in one particular situation, that doesn't mean that C, D, and E can't be the best orientation for others and their pets in that same situation or that our own orientations can't change under different circumstances. Over and over again we'll see how those who seek to ignore the paradoxes and champion one view as the only right one, or compromise the extremes into some behaviorally and emotionally neutral middle state, are the ones who get themselves into the most problems with their cats.

But isn't compromise the most effective, unemotional, objective resolution to the situation? It would be if we were dealing with an unemotional, objective problem. For example, if we're hungry and want to go out to eat with a friend, we may argue about whether Chinese or Italian food fits the bill, but eventually settle on the new French restaurant down the street. You want to buy an Alfa-Romeo, but your spouse wants a Continental sedan, so you compromise on a sporty Mustang convertible. As long as the primary goal is unemotional and simply involves satiating the appetite or providing transportation rather than enjoying a particular type of cuisine or driving a particular car, compromise works well.

Introduction

However, human feelings about cats tend to be so potent, so well defined and contradictory, that compromise offers no advantages. Consequently, we shouldn't be surprised to find that every population harbors both ardent cat lovers and passionate cat haters, what we call ailurophiles and ailurophobes. What does come as a surprise, however, and often a most painful shock to many cat owners is the cat's ability to elicit both emotional extremes within the *same* individual. While we marvel at the grace and agility a cat demonstrates as it leaps to bat falling leaves to the ground, we fight waves of anger and revulsion when the cat displays that same body language as it snags an unwary fledgling midflight. Instantly the graceful, alluring family pet becomes a demon, a psychopathic killer sadistically toying with a defenseless creature.

At such times our logical minds tell us to balance the two events, to see our cats as a compromise of the two behaviors; but our hearts seldom let us do that. The sight of my cat Maggie joyfully springing to swat every forsythia blossom served to her by gravity gives me a memory I'll always treasure; but the image of her eyes aglow with a satanic gleam as she deliberately and defiantly brings down a finch within minutes of consuming a bowlful of the best cat food money can buy appalls me so much that even the warmest memories can't possibly dilute my negative feelings in that instant.

Why does the cat, alone among domesticated species, elicit such powerful and paradoxical human emotions? Or perhaps more correctly, why do so many people allow cats to exert such power over their emotions? For better or worse—and the result is usually a mixture of both—*Felis domestica* provides humankind with an animate symbol of some of the most intimate and perplexing paradoxes inherent in nature. How we respond to the paradoxical cat, whether we adore and enjoy or fear and despise it, often tells us more about ourselves than about the feline species itself. Whereas most of us can accept our pack-oriented social dogs as symbols of fidelity and companionship—two positive and easily identifiable qualities we can count on our dogs to

Introduction

display—we frequently see cats as symbols of the inexplicable and unpredictable. Having so defined them, much of what they do surprises and even confuses us; and this, rather than our understanding, determines our relationships with our cats.

For example, imagine playing fetch with your faithful hound. Most people consider this a typical human/canine interaction. However, when a cat retrieves, we often consider the act extraordinary. Depending on one's orientation, the retrieving cat is either expressing its extreme intelligence by allowing itself to be trained or showing its incredible stupidity by permitting itself to be manipulated by humans into performing an uncharacteristic body-language display. Regardless of which view we embrace, we'll imbue it with a greater emotional charge than we would if faithful Fido performed a similar display.

Remember the old saw, there are three sides to every story—yours, mine, and the truth? That really holds true for feline body-language displays. For every cat behavior that creates often wildly divergent and opposing owner responses, we can find an ethologist's or animal behaviorist's view that claims to present the absolute, unemotional, objective explanation for that same display. In the case of the smart versus the stupid retrieving feline, the behaviorist might view the display as a reasonable and logical remnant of a genetically determined and refined predatory mechanism: Predators tend to survive if they consume their prey in a safe environment. Not only does this keep them from losing their food to scavengers, it keeps them from falling prey to other, larger predators. Consequently this retrieving behavior became incorporated into the wild feline's genetic pool; so it shouldn't surprise us when a domestic cat instinctively (or with minimal encouragement) returns real or pseudo kill (mice, birds, or toys) to places or people it considers safe.

This sounds reasonable, doesn't it? The behaviorist's "truth" isn't particularly problematic, and owners can use this view to enhance their individual human/feline relationships. Those who believe that their cats are extraordinarily intelligent can garnish that belief with an awareness that the display carries with it a genetic message passed from cat to cat for thousands of years. Any who previously viewed the body language as a sign of feline stupidity can now choose to see the display as a reflection of normal animal behavior.

However, sometimes the behaviorist's truth doesn't serve to enhance the relationship or reconcile opposing views of feline body language. Sometimes the behaviorists' logical and unemotional explanations serve only to make a bad situation worse or reduce a positive one to an objective void. Again, predation provides a fine example. I've read countless books and articles on predatory behavior and can easily accept the validity of most of the data and conclusions presented. However, all that knowledge does nothing to lessen the complex emotional response that always accompanies my discovery of a lifeless mouse or shrew on the braided rug outside my kitchen door. In fact, I can accept the revulsion only because it offers a tangible reminder that death is an unavoidable fact of life, a fact each of us must eventually confront on one level or another. But still, I hate to literally stumble over that ubiquitous unresolved paradox of life and death, hunter and hunted, villain and victim, at 6:00 A.M. as I wander out the door in my flannel nightgown carrying my watering can to the flowerbed. Because I'd rather avoid such heady lessons in my groggy early-morning state, the confrontation comes as a shock. I feel trapped and betrayed by Maggie for this assault on my senses.

No, I don't scream at her for being so wicked, nor weep over this senseless sacrifice of life—at least I try not to. On the other hand, I can't ignore what has occurred. Like so many owners I struggle to achieve a balance, a truth that works for Maggie and for me. But those damnable uncompromising paradoxes make me want to give up at times. The rich golden brown of the still-warm adolescent field mouse in my hand seems a poor trade-off for the joy I experience when Maggie pounces on petals and sunbeams. Wouldn't it be easier just to restrict her to the house?

Some people might shout, "Yes!" Others might yell just as loudly about the cat's instinctive need to hunt and its preference for freedom and independence. Logically, I could defend either position, even though they seem mutually exclusive. Emotionally, I'm a wreck.

In the pages ahead we're going to explore the varied human emotional responses to some common and typically feline body-language displays. Periodically some of the behaviorists' unemotional facts may come too close for comfort, suddenly revealing intimate personal beliefs we prefer to attribute to the feline "mystique." Sure, we all know what we *think* cats look like they're enjoying when they roll in catnip, but few of us actually come right out and say it. And when the scientists do say it, we feel trapped. Should we admit, "I knew that all along" and reveal a heretofore unacknowledged lascivious streak in ourselves and our cats? Or should we first feign ignorance, then horror? "Is *that* what she's pretending to do? Oh my God!" O cursed cat, stirring up such emotional dust kitties in the darkest corners of the human psyche!

As we unravel feline anatomy, physiology, and behavior and its relationship to human emotions and ultimately the bond between human and cat, we'll discover that we can offset every cat owner's lament with an equally powerful salute. For every feline behavior we view as brutally primitive or blatantly sexual, there's a complementary view of this same display as sublimely sophisticated or exquisitely sensual. The true or proper response for any given owner may be both, neither, or even immaterial. The only "truth" that holds any real meaning for us cat people is how our response affects our relationship with cats in general and our one special cat in particular. And while we may at times despise cats for taking us on this wild schizophrenic roller coaster ride between love and hate, good and evil, servitude and mastery, finicky discretion and indiscriminate independence, we can't deny the one gift the experience bestows on all humankind. By displaying the extremes, the cat presents us with the opportunity and choice to experience the entire range that lies between those extremes. Our human/feline relationships may border on the incomprehensible, but they invariably of-

Introduction

fer us that rare luxury of making responsible choices among paradoxical elements. When we lack an understanding of basic feline anatomy, physiology, and behavior and how our interpretations of these shape our relationships with our cats, it's easy to see the cat as an unfathomable mystery, the star of a magic show beyond our comprehension. But once we discover how these factors work together, the cat's magic becomes part of our own.

1

FAME AND INFAMY: THE CAT AND HISTORY

*D*ON'T touch Fluffykins while she's sleeping," chides Helen Dorchester as she tiptoes past the silver Persian napping on a satin pillow. "Isn't she the most regal animal you ever saw?" Helen glances at the clock: "Oh my, it's time to prepare Fluffy's seafood bisque."

Next door, Dick Lawrence heaves a rock at the stray black cat streaking from his yard with a bird in its mouth. "Get outta here, you damn devil!" he screams.

And a few blocks away Bob Klafen hands his wife a fuzzy kitten. "Well, we can't have any kids until we pay our school loans and save enough money for a down payment on a house, but we can still have our own little 'family.' If this one works out, we'll get another, maybe two."

Gods, devils, surrogate children: The domestic cat has played all these roles throughout history. Except for dogs, no other species of animal has enjoyed such intimate contact with the human race, and a study of the relationships between people and cats through the ages sheds light on the fascinating evolution of human responses to feline body-language displays. The story unfolds with all the beauty and startling twists of human evolution itself. If you took a basic high school biology course, you probably heard that "ontogeny recapitulates phylogeny." This somewhat intimidating phrase simply means that, to some extent, the development of an individual (its ontogeny) mimics the evolution of the species to which it belongs (its phylogeny). For example, if we accept that mammals, both humans and felines, descended from some sort of aquatic ancestors, we can see evidence of that evolution by examining the developing fetuses of either species. The early stages of both forms look distinctly fish- and amphibian-like before eventually assuming their familiar mammalian, then obviously human or feline, features. By studying the development from conception to death of an individual member of a given species we can glean a good deal of information about the history of the species itself. Conversely, knowledge of the species's evolution provides clues to the probable development of any one individual within that species. Taken together, these two different forms of history and evolution—the individual organism on the one hand and the collective species on the other—can give us a fairly complete understanding of the animal in question.

In this book we'll lay the foundation for such understanding and then use it as a doorway into the wonderful and little explored territory of interspecies body language and emotion. As we explore the mysterious and inscrutable cat and the sometimes seemingly unfathomable relationships that form between it and people, we'll obtain a clearer picture of both species, a picture that will help us get the most out of the cat-owning experience. Although most of this book will cover the "ontogeny" of human/feline interactions, the development of your special relationship with your special cat(s), we need to begin our exploration by going back to the earliest point in history, to when the two animals set up housekeeping together.

ALTERNATING CURRENTS OF THOUGHT

As we begin to review the history of human/feline interactions, one fact stands out: They clearly lack the consistency that has characterized peoples' relationships with dogs. Although cats have always entertained their devout champions and supporters, even when the majority of humankind associated them with the dark forces of Satan, they have never won consistent praise for particular traits the way faithful Fido has.

A brief survey of a wide range of cat-related literature demonstrates this lack of uniformity. Writers as diverse as Colette and Hemingway see cats as powerful symbols of the intimate and often contradictory needs of men and women, husbands and wives. Edgar Allan Poe and Guy de Maupassant curdle our blood with tales that take common feline characteristics and expand them to horrifying proportions, preying on our lack of understanding of normal cat behavior. Writings from such divergent authors as early Egyptian priests, fourteenth-century Christian monks, and the often bawdy social commentators such as Geoffrey Chaucer often speak of cats in sexual and sensual terms that set translators and interpreters at each others' throats. Conversely, authors such as the preserver of the legend of Dick Whittington's cat and Tennessee Williams write about cats that display almost doglike devotion.

Even though human beliefs regarding cats may be uniformly inconsistent, we can't escape the fact that they are also often uniformly passionate. Whatever people may believe about cats, chances are they'll believe it strongly. Although certain aspects of human/feline relationships do tend to recur, the historical cat, unlike the dog, cow, or horse, has triggered volatile and extreme emotions, ranging from deification to vilification.

What historical factors contributed to such wildly divergent views of the same basic animal? Have people changed so dramatically during the last five thousand years? Have cats changed all that much? Or is there something about cats that just naturally ignites human emotions?

To be sure, people have changed since 3500 B.C., but basic human biology and behavior have remained essentially the same. This is equally true for cats. The *Felis domestica* stalking a mouse in a suburban American basement looks and acts like the one that preyed on mice in ancient Egypt. Furthermore, the domestic cats described in ancient Egypt share

THE BODY LANGUAGE AND EMOTION OF CATS

the same characteristics with their prehistoric ancestor *Dinictis*. The agility, adaptability, and intelligence that enabled *Dinictis* to survive the centuries relatively unchanged still captivate cat owners today. Therefore, we must conclude that human beliefs and emotions regarding cats account for the tremendous inconsistency with which cats have been experienced in human society over the years.

The study of the relationships between people and animals reveals that awareness of and response to a particular species depends to a large extent on our ability to link it to our own experience. If the animal reflects a quality we recognize and like, we eagerly develop a relationship with it. For example, animals such as dogs, horses, cows, chickens, and ducks tend to be social, preferring to be among their own kind, a quality they share with humans. In addition these species also possess other qualities that we perceive as enhancing human existence: The horse serves as a predictable beast of burden, the dog as hunter, herder, and companion; the cow and fowl as reliable sources of food. Because we see all of these qualities in a positive light, selection and domestication progress with relative ease.

If we dislike certain qualities, we either try to ignore them or we attempt to eradicate the quality or the species, depending on how much the negative quality threatens us. When we feel threatened by certain rodents, snakes, roaches, and stinging or biting insects, we either try to stay out of their paths or spend a great deal of time and effort trying to wipe them off the face of the earth. If we feel that our meat should be leaner, our milk have more butterfat, our horses run faster, or our Persians be longer-haired, we initiate controlled programs designed to eliminate the undesirable genes and/or individuals.

However, even though we may tamper extensively with individual species within the animal kingdom, the fact remains that the human view toward most species has been fairly consistent throughout history—except for the cat. In spite of a tremendous amount of superficial genetic manipulation to alter its visible features (coat length, texture, and color; eye color; conformation or body shape), the cat's basic qual-

Fame and Infamy: The Cat and History

ities have remained relatively unchanged for centuries; however, human attitudes toward those qualities have fluctuated wildly. To understand why this phenomenon occurs, we're going to track human responses to four common feline displays:

- Nocturnal behavior.
- Territorial and asocial behavior.
- Mating and maternal behavior.
- Predatory behavior.

By contrasting human beliefs about these behaviors within the highly ailurophilic ancient Egyptian and the equally ailurophobic medieval European cultures, we can learn some fascinating lessons about the fundamental relationship between *Homo sapiens* and *Felis domestica*.

GODS AND DEMONS OF THE NIGHT

Remember the nursery rhyme about the little girl with the curl in the middle of her forehead? "When she was good, she was very, very good, but when she was bad, she was horrid." From ancient times philosophers have noted that anyone or anything that wins recognition from some quarters for being very, very good can quite easily elicit hatred and intolerance from others. Like yin and yang, "godlike" and "devillike" seem to form inseparable sides of the same coin. Strong human personalities (Socrates, Jesus, Hitler, and even Clint Eastwood) can evoke extreme love/hate emotions; many fashions (hairlength, hemlines) entice or anger depending on the era.

Because humans are a diurnal or light-active species, we've always been afraid of the dark, especially before electricity lighted our homes and streets. Therefore it's not surprising that the cat's ability to move freely in the dark and its apparent preference for a nocturnal life-style made a deep and lasting impression on our ancestors. Because the Egyptians recognized multiple gods, they could easily ascribe positive meanings to the cat's distinctly nonhuman behavior. Recall the flash of brilliant light emitted from the eyes of a night-stalking cat caught in the beam of your car's headlights. We can appreciate the Egyptian's

THE BODY LANGUAGE AND EMOTION OF CATS

assumptions that a creature with such ability represents something otherworldly. Because they didn't know about the highly reflective structure at the back of the cat's eyes (the tapetum), they easily assumed the cat was the *source* of that light. And because the dark-fearing Egyptians worshiped all light, this four-legged fur-covered "lamp" was worshiped too. In such a way the Egyptians took great comfort from the idea that the cat was a minireservoir of the sun god's great power, keeping watch over them during fear-filled nights; and out of gratitude, they made the cat a god.

Anyone who's looked closely at the feline eye can't help but notice how the pupil opens (dilates) or closes (constricts) in response to light. This fact didn't escape the notice of the early Egyptians, who related it to the waxing (dilation) and waning (constriction) of the moon, an association that further linked the cat with night. Once they made this association, they eventually transferred to the cat the moon's power to control the tides, the weather, crop plantings, and harvests.

In such ways the Egyptians took their only domesticated, available nocturnal species, imbued it with the qualities of both sun and moon, and made it a symbol of light in the darkness. Whether their cats lay quietly beside them or prowled the night, the Egyptians believed the cats were protecting them from the dangers of the dark.

During the Middle Ages people harbored just as many fears of the night as did the early Egyptians and perhaps even more. However, unlike the Egyptians, who had firmly established a multitheistic religion providing all sorts of protective gods and tangible symbols in the form of animals or idols, early Europeans got caught in the transition between pagan and monotheistic Christian beliefs. Because Christianity didn't permit its followers to recognize any "good" god save the One beyond human comprehension, night-fearing medieval Europeans found themselves in a bind. Their fears convinced them that if those otherworldly things going bump in the night couldn't be good, then they must be bad—a view supported by churchmen, who saw nothing inconsistent about recognizing an intangible God but a tangible Devil(s).

Fame and Infamy: The Cat and History

Into this world sauntered the only domestic nocturnal species; but instead of enjoying respect and deification, cats quickly became symbols of Satan himself. In this reversed view, cats became the target of human fear, an emotion that often leads to violence. Almost overnight, the destruction of a cat became synonymous with triumph over the Devil, the Prince of Darkness.

LIVING THE ASOCIAL LIFE

Another feline characteristic offering insight into the human/feline relationship is the cat's singularly asocial nature. All other domestic species, including humans, prefer the society of fellow creatures; we, our dogs, and our sheep function most happily when among our own kind; we birds of like feather flock together. If we can't, we'll flock with birds of just about any feather, gravitating toward a member of a different species simply to fight off loneliness.

Cats don't necessarily share this behavior. All felines are, by nature, asocial or solitary animals who prefer to be alone except when mating or raising their young. Because of this preference for the solitary life, the relationships between people and cats take on qualities unlike those that characterize the bond between us and any other domestic species. Unlike our bond with faithful Fido, to whom we are drawn by the similarites between us, the one we form with Sylvester depends on the *differences* between the two species. As with other feline displays, this behavior can lead to extreme human interpretations.

Once we form relationships with "birds of a feather," we often feel more comfortable exploring relationships with more alien creatures. This may explain why the ancient Egyptians, having gained confidence from their fixed and stable society, felt drawn toward novelty. And what could be more novel than the solitary, nongregarious cats? The cat's aloofness stimulated the ancient Egyptians to raise its exalted status still further. After all, a god would naturally avoid intimacy with common folk and would never consider coming, sitting, or staying just because some mere mortal instructed it to do so. The Egyptians believed that feline independence provided concrete proof that the cat-god *chose* to live with them, to serve them and their families; it didn't need them, it didn't depend on them for anything. Conversely the feline's asocial proclivities also provided a constant reminder that the cat could choose to abandon its host or withdraw its favors at any time. This awareness spurred them to anticipate their cat's every need lest they displease and alienate it.

The cat's asocial nature actually served the Egyptians' needs quite well. While they welcomed cats into their households, the distance between human and cat, which they considered a signal of the cat's superior nature, facilitated their devotion. This behavioral and often spatial gap reminded them daily that gods and mortals, night and day, men and women, were different, and never the twain would meet. In such a way the Egyptians relied on the asocial cat to help them accept and celebrate those aspects of their lives that they found most incomprehensible.

Medieval Europeans, on the other hand, were still struggling to get comfortable with monotheism and Christianity and longed for the security of predictable uniformity in their lives. Unlike the Egyptians, who evolved their civilization and feline-inclusive multideity worship over a period of approximately thirty-five hundred years, the early Europeans found themselves assaulted by domestic cats and Christians almost simultaneously. Unlike the Eastern cultures, they had no opportunity to worship the cat first, establish a safe distance from which to study its differences, and learn to accept it as beneficial.

Imagine a caravan of fervent preachers rolling into your town and joyfully proclaiming that salvation would come only to converts who owned computers and kept tarantulas as pets. Not only would these proselytizers be asking you to abandon religious beliefs that have supported your culture for centuries, they'd be asking you to accept these ominous-looking arachnids as harmless and necessary companions. Because cats were so familiar to those earliest Egyptian monks determined to Christianize Europe, they never stopped to consider what a pro-

Fame and Infamy: The Cat and History

found effect their totally different pets would have on their targeted converts. Essentially they were unwittingly thrusting a dramatic and powerful paradox onto the pagans: On the one hand they touted the virtues of a single God, but on the other hand, they brought with them the most unworldly godlike creatures the Europeans had ever seen. A legend tells of an early monk named Su Cat who traveled with his cats to convert the inhabitants of what would later become Ireland. His persuasive arguments regarding God's power were greatly enhanced by his (or was it his cats'?) ability to rid the region of the snakes that overran it. Su Cat's baptized name was Patrick, and many of the early churches founded in response to his teaching boasted cats carved from stone.

In Saint Patrick's case the cat survived the transition from multitheism. However, as time passed and the cats multiplied, the cats so far outnumbered the Egyptian monks in Europe that the latter could no longer neutralize the Europeans' fear of this species. As we've already noted, in their efforts to integrate a new religion and a new species, more often than not the Europeans saw the Church as very, very good and the cat as very, very bad. Because they were struggling to belong, to be what this new religion wanted them to be, anybody or anything alien became a terrifying threat. Obviously the aloof, asocial, noncongregational cat fell into this category, and the medieval mind perceived the asocial behavior as a flaunting of God's will.

In fact, according to early churchmen, cats acted a lot like the most dreaded outcast, Lucifer himself. The cat's refusal to obey like a dog signaled its inferior intelligence, stubbornness, and antagonism toward man and God alike. Its tendency to come and go as it pleased, independent of human desires, violated the law decreed by God Himself that all creatures should accept human dominion. In effect, the early Europeans said, "If you're not with (like) us, you're against us. If we're trying to be like Christ, you must be trying to be like Satan." As more and more clergy came from within the ranks of the Europeans, this attitude prevailed until the official view of the Church under

THE BODY LANGUAGE AND EMOTION OF CATS

Pope Innocent in 1484 empowered the Inquisition forces to burn all cats and cat lovers.

While it would be tempting to ailurophiles to side with the Egyptians and condemn these early Christians, we can't overlook the positive role the cat played in the transformation of a great number of multideity worshipers into a monotheistic society in a relatively short period of time. By projecting many of their pagan beliefs and existing fears on the cat and then destroying it, people rid themselves of emotional and philosophical burdens that would have otherwise greatly delayed their acceptance of monotheism and Christianity. Thus to some extent we may say that the cat became a scapegoat or a sort of sacrificial lamb, and in so doing remained godlike. Such is the glory and the steep price paid by those who dare to be different.

STAKING A CLAIM

Inextricably related to the cat's asocial nature is its strong sense of territoriality, its desire to establish, maintain, and protect a particular area. For many people territoriality is synonymous with possession, and oftentimes *exclusive* possession. And because human beliefs about possessing or being possessed by someone or something are often so intimate, such feline behavior was unlikely to escape notice.

As we would expect, the Egyptians and medieval Europeans held two completely different views of territoriality. Because the Egyptians wanted the cat to stay near, they rejoiced when its territorial displays included their households. The scratching and spraying of urine were viewed as tangible evidence the cat-god enjoyed the environment and was willing to protect it (and its inhabitants). Here again, what captivated the Egyptians was the cat's *choice* to include them in its territory, to count them among its possessions. However, even when the cat established a territory that excluded them, the Egyptians had little difficulty accepting the behavior. Their religious leaders and oracles often isolated themselves from human interactions, and such behavior was viewed as befitting the godlike as well as the god. So while they preferred that the cat honor them with its society, the fact that it chose not to wasn't seen as a feline deficiency.

At the other extreme the medieval Europeans saw the close attachment of cat and territory as a violation of God's and human will. That the cat would presume to possess anything was seen as evidence of its inherent wickedness, an assault to the blessed state of poverty and human dominion. To the devout who were demonstrating their piety and devotion by living the most spartan existence, the cat's bawdy brawls for mates and territories surely appeared demonic.

Those who believed God expected them to be able to control and dominate all creatures surely saw the cat's total disregard for human boundaries as a blatant affront to God and man alike. In fact, the association between claiming and protecting territory (that is, possession) and evil grew so strong that those suspected of such unacceptable behaviors were said to be "possessed" by the Devil and part of his territory or kingdom.

Unfortunately the medieval view of territoriality held that any person included in a cat's territory was obviously cursed or possessed by the devil, too, quite opposite the Egyptian view, which deemed such individuals the most blessed by the gods. This guilt by association led many Europeans to destroy cats rather than risk being accused of fraternizing with the Devil. At the height of the Inquisition, when cats and cat lovers were routinely being impaled, burned, and crucified, few were willing to be seen anywhere near a live cat.

ROMPING WITH THE WITCHES

Another peculiarly feline characteristic that catapults cats into fame or infamy is their tendency to relate to people the way kittens relate to their mothers (or queens). If we combine this display with the cat's natural grace and flexibility, we can easily comprehend how cats became symbols of femininity as well as motherhood. Every cat owner notices the way kittens knead when petted, spreading their front toes and pressing against one's lap or chest in a display identical to that

THE BODY LANGUAGE AND EMOTION OF CATS

used by the kitten to milk the queen's mammary gland during nursing.

The fact that both women and cats enjoy reputations for having intuitive powers reinforces the feline/female analogy. Sorceresses, oracles, and priestesses weren't uncommon in ancient civilizations, and their cats could not help sharing their supposedly clairvoyant capabilities. The ancient Chinese used cats to predict earthquakes and other natural disasters, a practice that has only recently come under sophisticated scientific scrutiny. Early sailors relied on their cats to foretell storms and other dangers in addition to protecting their cargoes from rodents.

While such analogies may seem remote from our contemporary beliefs, as recently as the Victorian era some people touted cats as the ideal role models for little girls, who should strive to imitate the cat's grace, poise, cleanliness, and mothering skills. A basic parenting guide of the day suggested that parents give each daughter a gentle female kitten to ensure that the child would learn the importance of good grooming and cleanliness at a tender age. When the cats matured and bore kittens, the girls would then learn the arts of patience, diligence, and attentiveness so useful to mothers. Armed with such knowledge, the young women could assume their rightful places in the world. That some maidens dared to question both such "teachers" and their lessons was illustrated by Alice's somewhat ambivalent attitude toward her cat Dinah in Lewis Carroll's *Alice in Wonderland*. Alice notes that Dinah is quite gentle, clean, and well behaved, but also an accomplished (mouse) killer. Once Alice tumbles down the rabbit hole, the semi-goodie-two-shoes role model Dinah gives way to the Cheshire Cat, an animal few would associate with a dutiful wife and mother!

Interestingly, while the Victorians embraced the feline qualities of grace, poise, motherhood, intuition, and cleanliness as exemplary for women, they totally ignored another feline characteristic that greatly influenced ancient Egyptian and medieval European attitudes toward cats. Among its peculiarly feminine feline characteristics the cat can count its unique ability among domestic creatures to experience sexual

Fame and Infamy: The Cat and History.

orgasm. While proper Victorians delicately ignored this critical preliminary of motherhood, it played a crucial role in the relationship between humans and cats in other societies.

As we might expect, the Egyptians and medieval Europeans embraced distinctly different views of the feminine characteristics of felines and the feline characteristics of females. Even though we all know that offspring don't result without the cooperation of both male and female, most cultures have more strongly ascribed fertility and reproduction to females and female symbols, probably because women actually carry and bear the children. Given this almost universal link, we progress fairly smoothly toward equating gracefulness with sensuality and sensual gracefulness as a sexual manifestation. When cat worship peaked in Egypt around 950 B.C., the cat-goddess Bastet, who was honored annually with a great festival in the city of Bubastis, combined all these feminine characteristics. While various historians, undoubtedly influenced by their different upbringings, described the festival as "an expression of great religious zeal," "a drunken orgy," or "a bit randy," they all agree that it celebrated the mental, emotional, and physical attributes of woman as symbolized by the cat.

At this point the history of the alliance between cat and woman separates into two distinct paths, which don't rejoin until much later—and with grave consequences for both feline and female. As the Egyptian civilization faded, the fledgling Christian religion allowed no room for cat-goddesses. Nevertheless some folks, the so-called pagans or barbarians, continued believing in multiple deities. Most pagans were nomadic herdsmen and hunters, whose womenfolk either sat home and waited for the men to return or dutifully dragged their meager belongings and children behind their husbands as they moved from camp to camp.

When these nomadic women discovered that cats could eliminate rodents, thereby facilitating food storage and the establishment of permanent homes, they welcomed the furry feline as an answer to their prayers. In this way cat worship enjoyed a renaissance, and not surprisingly, the Norse goddess of love, Freya, assumed a role similar to that of Bastet, including having the cat as her personal symbol. What began as a worship of feminine and feline virtues in an attempt to get the menfolk to settle down deteriorated into "bachanalian" and "lusty" celebrations once again. Freya and her cohorts appear in paintings of the era carousing all over Europe in her famous cat-drawn chariot. These "cat wagons" stopped to pick up passengers at "cat houses"— and we all know what kinds of people were associated with those places!

Meanwhile the medieval church was busy constructing a hierarchy whose power rested with the celibate male. While this turn of events would seem to run counter to what was occurring among the pagans, in reality the two groups exhibited similar needs. While pagan men, threatened by the emergence of a highly romantic and demonstrative female-dominated religion designed to turn them into farmers and merchants, wanted to preserve their nomadic life-styles, the early churchmen strove to preserve monotheism with the celibate male as the symbol of God's power on earth. Both groups understandably opposed Freya, her cats, and all they represented and invested a great deal of energy in trying to destroy them. If cats disappeared, perhaps the beliefs underlying the worship of Freya would vanish in their wake. To the Church's regret, this violent approach, far from eradicating felines from the earth, often turned them into innocent martyrs.

Because female behavior seemed so alien and threatening to their established and/or desired ways of life, pagan men and clergy joined forces in what turned out to be a wholesale slaughter of cats, not a few women, and an assortment of cat-loving men. While the movement didn't accomplish its ultimate goals, it did manage to obliterate the explicitly sexual cross-species symbolism between women and cats.

Ironically the two groups responsible for the slaughter—the crusading churchmen and the barbarians—eventually had to share the blame for introducing plague-carrying rats into Europe, rats that ultimately only the cats could control. Perhaps the Cheshire Cat smiles its inscrutable smile at that amusing bit of recorded poetic justice; but once

Fame and Infamy: The Cat and History

again we mustn't overlook the contradictions and paradoxes inherent in these historical accounts. Bear in mind that the historians in both periods were inevitably males reflecting the workings of a male political or religious structure from a male point of view. Consequently it would be easy to judge the stated or historical view of the cat's role in a particular society as inextricably linked to its men's attitudes toward women. While such may indeed be true, the fact remains that we can gain a larger appreciation for the cat's role if we see it as reflecting a particular society's orientation toward that which is different, inscrutable, or unknown, rather than feminine.

PREYING ON HUMAN FEARS

The feline body-language display that evokes the strongest human emotional response is that associated with the cat's skill as a hunter. Some historians go so far as to credit the cat for making civilization possible because its hunting ability reduced the prolific and doubly antagonistic rodent population and allowed the Egyptians to store grain safely. Before cats and humans coexisted, rats and mice not only consumed large quantities of grain, they also fouled whatever they didn't eat with their wastes.

Imagine yourself an early Egyptian already charmed by an elusive, sensuous nocturnal feline. As the cat responds to your companionship and the tidbits you leave to ensure its presence during the fearful night, it spends more and more time near you. Soon you begin to find dead rats and mice near the vessels you use to store your grain. Perhaps you hear a nighttime scuffle that convinces you this feline addition to your household not only keeps away nighttime demons but protects your food supply as well. Before long you realize that the grain that once fed your family (and the rodents) for a week now lasts two weeks. Thanks to your cat, you need devote less time to the raising and storing of food. While your cat patrols your fields and your storage bins, you can now pursue other projects: writing essays, painting pictures, or designing fine fabrics, jewelry, or pottery.

THE BODY LANGUAGE AND EMOTION OF CATS

According to some historians this scenario played itself out all over ancient Egypt. Given the Egyptians' positive perception of cats, their felines' ability to rid households of rodents placed them on an even higher pedestal of godliness. Moreover, the cat's efficient and enthusiastic hunting ability favorably impressed the Egyptians, who admired the way cats appeared to relish the hunt, a behavior they saw as evidence that cats enjoyed performing such a valuable service for humankind.

If the Egyptians believed that cats could do no wrong, most medieval Europeans thought they could do no right. Whereas an Egyptian observing a cat stalking prey would marvel at the creature's extreme patience, grace, and agility, a medieval European witnessing an identical display saw in it nothing but proof of the cat's malevolent, sneaky, and deceitful nature. The Egyptians welcomed the predatory cat into their homes as a devoted family protector, while the early Europeans banished it as the Devil's agent just waiting for the chance to suck the breath from babes and cast wicked spells on unprotected adults. To them the cat's enjoyment of the hunt provided irrefutable evidence that cats come from the bowels of Hell.

While it seems impossible that these fearful Europeans didn't notice cats preying on rodent pests as well as "innocent" songbirds, their fears forced them to ignore this obvious fact and condone the wholesale slaughter of cats throughout the Middle Ages. At the same time, the Church's attitude toward cats and cat-loving people intensified the persecutions. Ironically, but befitting the cat's paradoxical position in history, inhabitants of many monasteries and convents kept cats during the Middle Ages while simultaneously raising the hue and cry for their extermination in the outside world.

We can easily reconstruct the logic that led people during this era to their definition of the cat's predatory behavior as evil. As we noted earlier, if the cat's nocturnal behavior denoted deviltry, it followed that the cat must have evil intentions. Having defined the cat as a symbol, or perhaps even the embodiment, of Satan, people could hardly add, "On the other hand, cats do perform an admirable service by keeping

Fame and Infamy: The Cat and History

the rats under control." To do so would be akin to saying, "Aw well, the Devil has his good points, you know." Anyone even hinting at such heresy could quickly find himself or herself joining a dozen cats in painful crucifixion or burning.

We must realize that these early Christians maintained a strong image of Christ as the innocent victim, the sacrificial lamb. Those making the transition from multiple natural gods to the single God of Christianity needed such strong mental images to reinforce their new beliefs; and it didn't take a wild leap of the imagination to see the image of the innocent, defenseless Christ being stalked by His wicked enemies mirrored in the poor robin being mercilessly pursued by a cat that seemed to delight in the exercise.

Still, the cat's value as a rodent controller lingered in the background until the double standard promulgated at the monasteries and abbeys crept into the general population. As people came to prefer city life and the pursuit of various vocations, some chose to become shopkeepers and millers. Regardless of their degree of religious devotion, these folks knew they needed cats to keep the vermin away from their goods. Consequently, they concocted ingenious ruses that allowed them to maintain a relationship with a feline without running afoul of the Church or righteous neighbors. With some memory still of their former multiple gods, and battalions of fairies and spirits still an accepted part of their culture, they comfortably invented a goblin (or "kabouterje," "colfy," "gobelin," or "brownie," depending on the geographical location), an invisible spirit who protected their goods in exchange for a saucer of milk. Like cats, goblins were nocturnal and preyed on rats and mice, but medieval shopkeepers visited by goblins swore that they never saw any cats around their stores and would definitely kill them if they did. "Praise be to God, the goblins kill the rats *and* keep the cats away—all for a saucer of milk" was the usual explanation given.

Although most historians agree that the persecution of cats lessened to some degree when infested rats carrying the Black Plague inundated

Europe, they disagree about the source of those rats. Because such a strong link exists between an individual's orientation toward cats and his or her interpretation of cat-related events, we might be able to determine, by examining a historian's position on the issue, whether that historian is or was an ailurophile, an ailurophobe, or indifferent to the species. Some insist that the plague-carrying rats accompanied Asian barbarians as they overran the continent, while others claim that the Crusaders brought the pests back with them from the Holy Lands. In fact, both theories could be true, because the Black Plague appeared and decimated a large portion of the European population soon after both events.

This particular phase of feline/human history underscores the way cats can provoke irrational reactions in otherwise rational humans. One historical account ignores the barbarian hordes entirely, seeing the Crusaders' inadvertent introduction of infected rats to their homeland as a bizarre form of divine retribution against those who had so wrongly persecuted cats. According to this version, those who hated cats had no protection against the rats and succumbed to the hideous disease, leaving those who had befriended the cats to inherit the earth. Reading such accounts, we can hardly miss the thinly veiled implication that the religious leaders who persecuted the cat sowed the seeds of their own destruction and that those who remained true to their feline companions won salvation.

However, such a view can't wave away the fact that hordes of barbarians *were* also wreaking havoc on this same population at this time. Nor can it entirely dismiss the position that the diseased rats might surely have accompanied the savage and unwashed barbarians. Perhaps we would be wise to take the middle ground, admitting that cats did play a role during the plague years and that despite the size of that role, the history of the period includes more than a simplistic battle between the Church and the cat, with the cat ultimately prevailing. Unfortunately, when it comes to cats, the middle ground usually turns to quicksand as our human emotions polarize around this predatory mystery.

Fame and Infamy: The Cat and History

THE FLEXIBLE FELINE

Having seen how some people reacted to feline characteristics during two extremely different epochs, we have obtained parameters within which we can measure the cat's current relationship to humankind. If you were to spend an afternoon at a contemporary cat show, you might go away convinced that some of those ancient Egyptian cat worshipers have been reincarnated in the form of cat fanciers of America, whereas a visit to a clandestine arena where handlers of illegal fighting dogs bait the combatants with live cats might make you feel you've been hurled back into the Dark Ages.

While such pockets of deification and vilification still exist, they certainly aren't the norm. The majority of human/feline relationships in Western civilization fall somewhere between these two poles, even though any one cat owner may vacillate wildly between them from time to time. When researchers first identified leukemia as a viral disease in cats, hundreds of owners blindly rushed to have their pets euthanized, although there was (and is) no evidence that humans can catch the disease from cats. A recent report in a veterinary journal described a case of bubonic plague in a cat, and even those of us who know that the disease crops up in this country couldn't help but cringe at the implications for feline and human populations. If authorities diagnose a rabid cat in a heavily populated suburb, letters demanding feline licensing and control soon flood the desks of health officials, politicians, and newspaper editors. One day my verbal urgings bring Maggie instantly to my lap, kneading contentedly as I pet her, purring out her devotion and love. The next day she responds to that same invitation as though I'd suggested something obscene. "Ungrateful wretch," I grumble, "accidentally" dropping a magazine within inches of where she smugly grooms herself as though I weren't even there. However, even though cats can still evoke the full range of human emotions, most owners want to maintain a harmonious balance of only the best.

With its rich and volatile history, the cat seems ideally poised now to enter into a solid bond with its human hosts. Unlike other ages,

THE BODY LANGUAGE AND EMOTION OF CATS

we're not bound by stereotypes that would pressure us to accept one radical feline orientation or another. While the lack of extremes may leave us without strong guidelines regarding the "right" kind of relationship to build or exactly how to go about building it, polar positions usually preclude the formation of unique liaisons based on individual needs. Without the uniform feline-related behavioral codes that standardized and stereotyped the actions of ancient Egyptians and medieval Europeans, we can take advantage of our newfound freedom to develop our own particular relationships with our surprisingly flexible cats. While modern times may not go down in history as an age of feline enlightenment, the potential certainly exists for this to turn into an age in which a wide variety of rewarding relationships became possible.

From god to devil to flexible feline, we've briefly traced the human/feline relationship over a five-thousand-year spectrum. As we now move to a detailed look at the physical and psychological makeup of today's cat, bear in mind that these historical feelings still influence us today. Surely all cat owners have relished the special peace that comes from a cat's presence on the bed during a long night when sleep won't come. Surely we've all caught our cats staring at us—or is it *through* us?—with otherworldly expressions on their faces. Within each evolving relationship between human and cat lurks a potential worshiper and a potential destroyer, ready to unite or clash with a feline god or devil. Now that we know that these potentials exist, and how they were manifested in the past, let's focus on achieving the unique and intimate balance that will ensure the formation of a lasting bond.

FELINE BODYWORKS:
THE PHYSICAL CAT

IGARO, no!!" screams Debbie Barclay as she hurls a pillow at the white Persian casually lounging in the comfortable nest he'd constructed for himself in Debbie's new black wool sweater. Figaro nonchalantly watches the pillow sail over his head, then he leans slightly to one side, digs his front claws into the sweater, and aims a hind foot at an invisible flea scurrying across his shoulder. Tufts of downy white fur waft upward, then settle on the dark sweater.

"You spiteful beast!" shrieks Debbie as she attempts to remove the hairs. "You did that on purpose!"

This typical exchange of human and feline body-language displays contains a rich array of feline anatomical components. When Figaro exhibits the relatively simple scratch reflex to discourage a flea, no less than nineteen different muscles must contract simultaneously and specifically to get the job done. Moreover, as the flea wanders across Figaro's dorsal (top) or ventral (bottom) midline, the Persian will automatically switch legs to attack the pest in its new location. If we add all the muscles necessary to keep Debbie's feline companion from toppling over while he scratches, the skeleton and blood vessels necessary to coordinate all these activities, Figaro's simple scratch becomes an awe-inspiring anatomical and physiological feat. If we also consider the subjective brain functions involved (that is, why Figaro pursued this flea and not the one near his left elbow), the complexity becomes mind-boggling.

In the next two chapters we're going to examine the anatomical and physiological foundations of the common feline body-language displays that exert the greatest influence on the formation of a stable human/feline bond. Although a thorough dissection of all the anatomy and physiology that contributes to the cat's unique relationship with humans would require volumes, we're going to limit our discussion to a few of the most influential structures. Once we understand what's going on in these feline bodyworks, we can turn our attention to understanding how the resultant displays affect our responses to our pets' behavior. Even though Debbie reacts emotionally to Figaro's clawing, ascribing to the cat some diabolically evil intentions, the fact remains that Figaro's display occurs as a logical and predictable consequence of normal cat anatomy and physiology. Unless Debbie pauses to ponder this fact, she runs the risk of introducing some erroneous and even harmful elements into her relationship with her pet.

IT'S INTEGUMENTARY, MY DEAR WATSON

George and Mary Caldicott named their domestic shorthair Watson because his glossy black coat, white bib, and somber, attentive expression gave him the aura of a proper English gentleman. While they adore the way Watson rubs his head on their ankles as he weaves between their legs, his constant attempts to use George's favorite chair as a scratching post threatens to undermine an otherwise stable relationship.

Like so many cat owners, Mary and George marvel at Watson's virtues one minute and condemn his vices the next, never realizing that both often utilize different parts of the same anatomical system. Many familiar feline behaviors involve the integumentary system, which is composed of fur, skin, whiskers, scent glands, and claws, among other structures. As we consider each of these components and its effect as the cat interacts with its environment, we'll soon see that whatever

Feline Bodyworks: The Physical Cat

a cat isn't prepared to subdue or outwit, it's probably prepared to charm or confuse.

One of the first things we notice about an animal is the color of its fur. From a practical, survival point of view, the most beneficial fur color for a wild animal protects it under the widest range of circumstances it might confront in its environment. Given the cat's nocturnal tendencies and predilection for rodents and other small game, we can see how a variable brown striped pattern offers the maximum advantage. While modern cat fanciers may attribute all sorts of aesthetic and show advantages to particular stripe width, color, and sequence, the basic tabby pattern mimics the shadows created when light falls upon vegetation. Once we become aware of it, we can appreciate how this natural pattern enables the cat to blend into its surroundings. One veterinary technician named her cat Autumn, an unlikely name for a kitten born in June; but the name perfectly described the banded patterns of dark and lighter browns and oranges that rendered the kitten virtually invisible in a pile of leaves. The shadowy stripes of my own common tabby Maggie make her as difficult to see in the sunlit woods as when that same environment is moonlit; and in the half-light of dawn or dusk, her variable brown coloration enables her literally to fade into the background.

Not only does such coloration and patterning enable cats to creep as close as possible to their prey without being seen, it also protects them from falling victim to other predators. Consequently, we may assume that those early domesticated cats people encouraged to hunt shared a common color and pattern that best promoted their survival in their particular environments.

As domestication progressed and cats became more numerous, the novelty shifted from owning *any* cat to owning a particular *kind* of cat. Instead of being attracted to the norm, people sought out the exceptions that form the foundations of many present-day breeds. The extreme colorations that became so popular arose less out of any real uniqueness than out of different manifestations of existing anatomical and physiological potentials. Basically the color of a cat's coat depends on the relative presence of one pigment, melanin, and its orange variants. A lack of melanin in the hair makes a cat appear white; a lot of it makes a cat look dark brown or even black. In terms of camouflage, cats of either extreme are more vulnerable because they no longer enjoy the classic protective coloration and patterning. Not only would they lose an edge when stalking prey, they could also become more highly visible targets themselves. Not surprisingly, both black and white cats were rare in the natural population, and as with most rare objects, humans attached a great emotional charge to them. Depending on the period of history and the particular human culture, both white and black cats have been deified and vilified. While there are still people who adore or detest black or white cats, as feline genetic manipulation has become more common, a dizzying array of coloration and patterning has emerged.

For example, the unique pigmentary distribution of the modern Siamese can be traced to a genetic mutation first documented in Thailand, where early manuscripts speak of white cats with black paws, tail, muzzle, and ears, and blue eyes. It was as if the pigment that normally would have been spread over the cat's entire body had been concentrated in its extremities, creating a genetic paradox—an albino (pigment-lacking) cat with too much pigment in some areas.

Other records indicate that a similar mutation occurred in several areas simultaneously, perhaps as the result of viral infections or environmental changes that assaulted the developing young but not the queen during pregnancy. While this is certainly possible, the coloration afforded no camouflage whatsoever, and those individuals not nurtured in protected human environments undoubtedly perished.

While we'll probably never know for sure what happened, the fact that the creamy color of today's Siamese is actually a bleached-out version of the ancient tabby color and pattern can't be ignored. One of my professors routinely proved this point by placing Siamese under bright surgical lights, where even the most uniformly colored revealed faint tabby stripes.

Feline Bodyworks: The Physical Cat

Moreover, while normally the rules of camouflage dictate that seasonal color changes produce lighter (whiter) coats in winter to offer protection in snow-covered environments, the Siamese mutation creates just the opposite. Siamese cats allowed outdoors in colder climates tend to be much darker overall than those of the same breed kept indoors. (Soviet scientists discovered they could maintain the dark points—face, feet, and tail coloration—in outdoor animals only by clothing them in wool sweaters. Otherwise the cats lost the light body color and assumed a more or less uniform darker hue.)

THE LONG AND SHORT OF A HAIRY DILEMMA

After we note the color of the cat's coat, we usually turn our attention to the length of that coat. If the cat functions as a predator in a temperate climate, short hair confers a number of survival advantages:

- It provides maximum individual temperature control in the widest temperature range.
- It can be easily cleaned.
- It interferes minimally with movement and function.

Because of variations in the texture and layering of cat fur, feline coats make efficient insulators, maintaining precious body heat in cold weather and resisting and radiating heat when temperatures rise. The cat's coat is composed of a series of layers, the density of which traditionally reflects the animal's genetic makeup, which in turn reflects its needs in a particular environment. Wild cats living in harsher, colder climates have thicker coats and undercoats than their tropical counterparts, who may appear to have minimal coat and no undercoat at all. Compare the smooth, glossy coat of the black leopard with the layered look of the snow leopard. The former's coat appears to be almost sprayed on the cat's body, whereas the latter looks as though it were peering out at its world from within a pile of woolly blankets.

When natural selection governs coat characteristics, animals inhabiting colder climates produce longer and denser coats, which provide numerous gaps that trap air and produce a maximum insulating effect. When fur-bearing animals get "goose bumps," clusters of hair are elevated, thereby increasing the air spaces and improving the insulating effect even more. (Because humans lack sufficient hair covering, this same reaction only increases the amount of skin exposed to the air, and we wind up feeling cooler, not warmer.)

Longer hairs also tend to angle more sharply away from the animal's spine. Compare a sleek Siamese and a fluffy Persian; the hair of the former flows almost parallel to the spine, whereas the Persian's coat angles toward the ground. While we don't tend to consider either breed more suitably adapted to a particular environment, wild-cat coat length strongly depends on the environment. While the shorter, smoother hairs aligned parallel to the spine permit maximum external air/body contact, minimal wind resistance, and therefore maximum cooling, the longer-angled hairs with their dense undercoats create the opposite effect, resisting external air currents and preserving body heat.

Short coats also require the least maintenance. Basic feline hair anatomy guarantees that all cats will shed, and basic feline behavior decrees they will groom. Unlike humans, all of whose hair grows at the same rate, feline hair grows at different rates, with new hairs constantly erupting as old ones die. Although this process occurs to some extent year-round, major shedding and regrowth occur in the spring and fall. Those who must daily groom an uncooperative Himalayan might find such an anatomical adaptation a pain in the neck, but it serves wild felines quite well, always providing the animal with some sort of protective skin covering as well as camouflage. The two shedding seasons reflect those times when the density of the coat requires the most alterations: the loss of the heavy undercoat in the spring and the replacement of the summer coat with a thicker one in the fall.

Ask any longhaired-cat owner to name their cat's most troublesome problem and they're apt to groan, "Hairballs!" Medications and magic cures for hairballs abound, and vigilant owners often use more than one. Does this mean that feline grooming behavior resulted from another genetic mutation perpetuated by foolish humans? Not in the least; in the wild predatory cat, grooming is essential to good health and

Feline Bodyworks: The Physical Cat

survival. When a predator makes a kill, it's bound to soil itself with blood and debris. Not only do such substances attract microorganisms and insects, which may be sources or carriers of infection, their odors may help other predators locate the cat. Consequently the cat that keeps itself clean eliminates two immediate problems associated with predation. Grooming also helps the cat eliminate external parasites, such as fleas, lice, and ticks, which it may pick up from its prey or during interactions with other felines.

Shorthaired cats can groom themselves more quickly and thoroughly than their longhaired relatives. Furthermore, because all cats constantly shed loose (dead) hairs, the shorthaired cat obviously swallows less hair during the grooming process and therefore develops fewer of the problems associated with hairball formation. Again, we can see that all these qualities of the shorter coat offer distinct survival advantages. However, we can also see how humans who took over the care and feeding of cats might find the less common, longer, and silkier coats more appealing.

Finally, the short-haired coat offers minimal interference when cats hunt, mate, and nurse their young. When the cat stalks prey low to the ground through thickets, brambles, and other hostile vegetation known to harbor rodents, such a coat offers minimal resistance. The short-haired felines also avoid two reproduction-related problems that often plague their long-haired counterparts (and their owners). It's not uncommon for longhaired tomcats to accumulate a ring of hair around the base of the penis. While this causes no harm or pain, it does interfere with mating. Second, even though much of their undercoats will shed by the time kittens are born in the spring, long-haired queens offer their young the added challenge and energy drain of having to work their way through this extra fur to get at the nipples. Although knowledgeable breeders routinely deal with both of these problems, these would clearly work against the long-haired cat's survival in the wild.

Many times owners obtain long-haired cats because of their striking appearance, only to discover that their own life-styles or personalities or the cat's temperament make it impossible to groom the animal sufficiently to maintain a healthy coat and skin. Because of the coat's variable growth rate, loosened and dead hairs become easily entrapped, and mats appear as if by magic. Once matting begins, grooming further tries the patience of both cat and owner, and a vicious cycle is set into motion. The mats not only ensnare debris, such as dirt, bits of twigs and leaves, food, stool, and urine, they also block the removal of the millions of skin cells routinely sloughed during grooming. Dry, flakey patches appear under the mats, which often irritate the animal, causing it to scratch or lap and tug furiously at the mat in an effort to remove it and get to the skin beneath. Often such licking and scratching only increases the amount of matting; if the cat succeeds in removing the mat, more often than not it will keep the area beneath raw with its continued licking. In the worst of all cases, flies attracted by the filth lay their eggs among the mats and the cat becomes an unwilling host to a population of maggots. While the image of a severely debilitated cat crawling with maggots may sound like something right out of the world's worst horror movie, for most veterinarians such matting-related problems are part of the summer routine.

What makes these problems doubly tragic is that they're totally avoidable as long as owners are willing to put aside their emotions and some mythology. If you can't properly groom your long-haired cat and want to keep it, you have two options:

· Have the cat professionally groomed on a regular schedule.
· Have the cat shaved down periodically.

The first option is more expensive, but the latter can carry a higher emotional price. Like long-haired dogs, long-haired cats generate their own unique mythology, which includes the "dangers" of clipping hair. To be sure, if the cat is shaved down when the bulk of the hair is relatively inactive, it may take longer to grow in; and depending on the cat's genetic makeup, the new coat may be a different color. However, these are simply cosmetic considerations, especially when com-

Feline Bodyworks: The Physical Cat

pared with the obvious health threat posed by the matted coat. Viewed in terms of "What's best for my cat and me," rather than breed standards or what Aunt Harriet says, many long-haired-cat owners have greatly enhanced their relationships with their pets by choosing to have them shaved down.

While this becomes an annual event for some owners, many owners, such as Debbie Barclay, discover they can establish a successful grooming program once the cat is shaved. When a brutal work schedule left Debbie with little time to groom Figaro for over a month, the cat looked like a four-legged mat. Debbie had the Persian shaved down to remove all the mats and then instituted a program to make grooming a more pleasant process for both her and her cat. When he returned home from the groomer's sans hair, Debbie set aside time each day simply to stroke her cat, first with her bare hands, and then with a soft cloth. Over a period of weeks as Figaro's hair grew in, she added brush and comb to the routine. In such a way the previously antagonistic grooming sessions gave way to ones that are essentially an extension of petting and a source of pleasure to owner and cat alike. By the time Figaro's coat had grown in completely, a valuable and healthy physical and emotional dimension had been added to the bond between Debbie and her cat.

THE UNDERCOVER STORY

If we could peek beneath the cat's fur and see the naked feline skin, we would detect two more alluring and elegant adaptations. Because much of the ritual fighting among intact males involves swatting and cuffing around the face, male cats develop protective pads of thickened skin on either side of the face called *shields*. These give toms their characteristic square-headed appearance and make them look much larger than females.

While the skin comprising the shields doesn't move very easily, that in the neck area does. Why didn't nature protect the readily accessible and vital jugular veins, trachea (wind pipe), and crucial nerves and arteries in this area with a similar thick and immovable covering? If combat were the only survival skill the species needed, it might have. However, thicker but looser skin in the neck area facilitates two behaviors critical to the survival of the species: transporting the young and mating.

Queens characteristically move their kittens by gripping the scruffs of their necks. As soon as the kitten feels its mother's mouth clamping down on the scruff, it becomes immobile and passive, its tail drawn upward against its abdomen. The survival advantage of being able to move such a compact feline package is obvious. Not only are there no cries to attract predators, the kittens' closely held limbs and tail neither flail about impeding the queen's progress nor are they vulnerable to damage from environmental obstacles the queen might need to negotiate.

When kittens mature, the thickened loose skin remains but benefits the adult in a completely different way. The mating male grasps the female in this area with his teeth. Now, however, the grip doesn't induce the female to curl her tail under her body but rather to draw it to one side in order to expose her vaginal opening. Consequently, the old immobility and the new tail orientation facilitate breeding.

An awareness of these anatomical adaptations and their behavioral implications can eliminate potentially negative physical and emotional human-feline interactions. Although it is possible to restrain a cat by holding it up by the scruff of the neck the way a queen holds her kittens, inexperienced people attempting such grips on adult cats court disaster. More than one person has been slashed by lightning-fast razor-sharp rear claws. Furthermore, adult cats who normally wouldn't find themselves suspended like young kittens might injure themselves if they panic and struggle without adequate rear-end support. Consequently, while cats can be restrained in this manner, it should only be used when necessary and only by those trained in the proper technique.

Feline Bodyworks: The Physical Cat

THE SCENT OF VICTORY, THE SWEET SMELL OF LOVE

When Watson lovingly rubs his face on the Caldicotts' ankles, he employs yet another integumentary structure, the scent glands. Primary feline scent glands are located at the corners of the lips, between the eyes and ears, and around the rectum. Befitting their dual and paradoxical nature, cats use two forms of marking; and, befitting our relationships with our cats, we tend to adore one form and despise the other.

Who can deny the pleasant feelings aroused by a favorite feline sweeping its head across our bodies with a blissful smile on its face? That's exactly the sort of behavior that makes the cat so endearing. However, if we trace the path of feline contact with human arm, leg or chest, we discover it often involves an arc stretching from the corner of the cat's lips to the top of its head, a graceful motion that applies pressure to the scent glands in this area and stimulates the release of identifying feline secretions imperceptible to humans.

"You mean Watson's caresses are no different than his spraying urine?" gasps the obviously distressed Mary Caldicott. Yes and no. Like spraying, face rubbing is a kind of marking behavior. However, it's not the same kind of marking behavior, because face rubbing seems to be a more passive display, the cat's way of touching bases with those objects or individuals in its environment whose presence it considers nonthreatening and beneficial. For example, my brother used to lie on the couch and idly stroke his cat while reading until she practically stuffed her whole head into his hand. On her way to the couch to share this nightly interaction, the cat would rub her face against the corner of the coffee table nearby; over the years she neatly rounded that once square projection. Because of the table's close proximity to the couch, my brother, and this satisfying interaction, the table itself took on the scent she defined as "something special."

Another example of this type of marking involved a friend I once astounded when I gently chided her for feeding her already overweight cat ice cream. The owner blushed and stammered, sure that I must be

THE BODY LANGUAGE AND EMOTION OF CATS

some sort of perceptual genius to know she hadn't heeded my recommendation to eliminate this daily treat from her pet's diet. Actually, I'd made a most elementary deduction. The no-ice-cream diet had supposedly gone into effect several months earlier, and my friend kept an immaculate house. Therefore the telltale cat-high smudge I'd noted on the refrigerator door could be nothing but the result of recent feline face rubbing. While I couldn't be sure the rubbing behavior didn't persist long after the owner ceased the ice cream treats, the cat's continued weight problem indicated the owner had yet to abandon the practice.

Most humans find such behavior acceptable because feline substances spread in this manner don't negatively trigger their own senses. Obviously if they did, few people would tolerate the behavior. And indeed, when cats activate more potent marking systems to protect their territories from threats, most humans turn up their noses in disgust. Although urine is the most common scent source for this type of marking, glands in the rectal area can also come into play. These secrete a substance whose long-lasting and pervasive obnoxious odor reminds us of skunks, or worse.

The cat's defensive territorial marking naturally results from its asocial or solitary nature. Because fighting expends a lot of energy, it only confers survival benefits when the results exceed this loss. Although any species profits from strong males who succeed in driving off competitors and win the right to mate, and from fighting queens who protect their young, the time and energy spent remaining on constant lookout for an attack from intruders would leave little time for anything else. Therefore, a system that permits the asocial animal to label its territory serves two purposes. First, it limits how much area the cat must actively protect; second, it warns other cats to stay away from that particular area.

Another scent gland, the tail gland, becomes particularly active in breeding males. Because males can get so totally engrossed in mating that they abandon all but the most cursory grooming, many of them,

Feline Bodyworks: The Physical Cat

especially the light-colored ones, develop what looks like a grease smudge near the bases of their tails. We know that this gland is somehow related to sexual activity because castration stops the secretion; however, we don't know for sure whether males use the secretion primarily to attract females or to discourage other males.

ON LITTLE CAT FEET

When Carl Sandburg equated fog with cats' feet, he only touched on one characteristic of the multipurpose feline foot. To be sure, the soft, spongy pads, coupled with their fully retractable claws, enable cats to move almost soundlessly, a distinct advantage for a predator stalking prey equipped with excellent hearing. Like the fog, the foot pads also deposit their own form of "dew," because their rich capillary blood supply functions as an efficient temperature regulatory device. When the cat's body temperature rises, countless tiny vessels in the pads dilate and dissipate heat. If your cat tends to get excited during veterinary examinations, you can observe the stainless-steel examination table to see evidence of this phenomenon at work. As the heat dissipates, moist footprints often form on the cooler metal surface.

Although this method serves as an excellent mechanism to rid the body of excess heat, the rich blood supply and permeable pad covering create a two-way transport system that can be problematic. Because cats are highly sensitive to certain substances found in some fertilizers, insecticides, paints, and automotive products, among others, it's possible for them to absorb toxic amounts of these substances through their feet. As if being hypersensitive and walking on sponges weren't bad enough, grooming further facilitates this process. Back in the modern Dark Ages when DDT was freely splashed about, numerous cats made their post-DDT-spraying foray across the yard and through the fields, rubbed posts and doorjambs to reestablish their territories, then groomed themselves as befitted their fastidious nature. In such a perfectly normal way, they signed their own death warrants. Similarly many thrifty cat owners who used wood ashes instead of sand on their icy sidewalks

The Body Language and Emotion of Cats

and driveways found that the resultant veterinary bills and mental anguish accrued when the cat fell ill from ingesting this material quickly canceled out any savings. Although public awareness and tougher regulations have eliminated many harmful chemicals from our daily lives, it pays to limit your cat's activity when any suspicious substances are being used until you can verify their safety.

Even though cat feet don't exhibit the characteristically whirled patterns that make human fingerprints so distinctive, they do possess scent glands, which serve the same purpose. Can you recall times when your cat put its front paws on a door or piece of furniture (or you) as though it were going to scratch, but it stretched and slid its feed down or across the object instead? This constitutes another form of passive marking of the "This is mine/special" rather than the "Go away!" variety. Owners of young kittens can condition their pets to scratching posts by gently rubbing their front paws on the fabric daily. In such a way the scent labels the post as the kittens', and instinct compels it to return to refresh the mark and reinforce the claim daily.

Clues about Claws

The last integumentary structure we're going to consider causes some sharp problems for many cat owners: the claws. At the base of each nail lies a blood vessel and nerve, which are encased in a series of modified skin layers arranged not unlike the outer layers of an onion. Anatomically the front and rear claws are identical, including muscle attachments that permit the cat to extend or retract them at will.

An exquisite anatomical adaptation, the retractable claw beautifully fulfills the needs of a nocturnal predator whose rodent prey has keen ears. We already noted how the cat can move almost soundlessly with retracted claws balancing its weight on its spongy footpads. To appreciate the value of this, we need only recall the sound of a dog and a cat walking across a hardwood floor. Unless canine nails are trimmed or worn down, they create a clearly recognizable clacking compared with the feline's almost silent tread. Dogs with exceptionally short nails

Feline Bodyworks: The Physical Cat

may mimic the cat's silent stalking, but must sacrifice any holding or defensive capabilities of their claws to do so. Not so the cat, who easily maintains its well-sharpened claws by concealing them within protective skin folds until needed.

Once we realize that feline claws are layered with the oldest tissue exposed and the newest closest to the blood and nerve supply, we can see that a proper feline pedicure involves more than the cat's simply sharpening the points of its claws. Periodically the outermost layer must be removed, otherwise the layers will continue to build up. Unfortunately this anatomical reality often escapes the attention of those who manufacture or purchase scratching posts. In order to remove the outer layer, the cat needs a relatively loosely woven material with periodic strong vertical and horizontal fibers. This fiber arrangement enables the cat to dig in the claws up to the base and pull downward, thereby peeling off the outer layer. If you ever find shredded nail "skeletons" at the base of a scratching post, you know the cat has been using the post for its intended purpose. Unfortunately, unknowledgeable owners will sometimes see these sloughed nails and assume the scratching post is too rough, breaking the claws instead of sharpening them.

What about cats who scratch on trees and doorframes? These cats are grooming, marking, or both. Although the newly exposed claw is razor sharp, the cat must periodically sharpen it to maintain its point; and hard surfaces make ideal sharpeners. In addition, the scratches made in these surfaces serve as visual warnings to others that enter this area. Maggie never made any attempt to claw in the house until I began letting the dogs indoors more during a particularly cold spell. Then she began clawing the doorway to the room that contained her favorite sleeping area and the ladder that led to her special loft. The less time the dogs spend in the house, the less she scratches these areas; all actual nail grooming still takes place outdoors.

While all of these physical expressions of behavioral patterns may be quite apparent in wild cats, the average domestic cat inhabits an environment where infinite variations can arise. For example, Debbie Barclay's Figaro may use the edge of the couch to remove any loose outer nail coverings, then saunter over to scratch his carpeted scratching post to sharpen his other claws. Having completed his pedicure, he may then rake his claws across the back of the velvet chair by the front window to remind the Siamese next door to "Keep out!" Unless we understand the anatomical, physiological, and behavioral components of clawing, such perfectly logical and normal displays could precipitate a wide range of emotional responses.

Is there such a thing as the "ideal" covering for a scratching post? Probably not, although understanding what the cat is communicating via the behavior often makes it possible to select a material that complements rather than antagonizes the behavior. I asked a long-established upholsterer what kinds of fabric he recommends to customers whose furnishings had been ravaged by cats. With a twinkle in his eye he showed me swatches of synthetics he *doesn't* recommend. These loosely woven fabrics often have a sprayed backing that makes the material stronger. While this provides sufficient support to withstand normal human wear and tear, it yields readily to claws. Similarly, the stronger fibers that periodically reinforce such fabrics make them deliciously inviting to cats. "My eldest son probably owes his college education to Herculon," said the upholsterer with a chuckle.

While such fabrics may be the *worst* coverings for furnishings (a smooth tightly woven, uniform fabric is the best) they can make marvelous coverings for scratching posts. How well I remember the back corner of my Herculon-covered couch; in terms of my cats' needs, it eventually provided everything. The remaining reinforced fibers bespoke the fabric's ability to remove the devitalized outer nail layers; the gouges in the exposed wooden frame attested to its function as a sharpener; the dangling shredded weaker fibers certainly provided a strong visual cue to other cats. None of these benefits were available from the dense carpeting that covered their own scratching post. When we built the post, I chose a material I thought would last. Well it did, but not because it was so strong; it lasted because it was so unsuitable the cats refused to use it.

Many owners discover that two scratching posts are better than one. They cover one with the aforementioned loosely woven synthetic upholstery material, which is periodically replaced. The second post is simply an eighteen-inch-long, six-to-eight-inch-diameter log nailed to a sturdy base, or split and nailed to the wall in an accessible location. Between these two materials, the majority of cats are able to satisfy their claw-related physical and behavioral needs nondestructively.

Before we leave the topic of claw anatomy and physiology, we need to consider how a cat's age affects its ability to maintain its claws. Sloughing, sharpening, and marking displays all require a fair amount of physical agility and strength, particularly in the hindquarters on which the cat must balance while performing these activities. Even though the aging house cat may feel neither desire nor necessity to sharpen its claws or mark its territory, the nails will continue to grow and the outer layers accumulate. Some older cats (as well as those who won't use available scratching posts for one reason or another) may attempt to chew the outer nail covering off. Others seem oblivious to the ever-expanding and lengthening nail, until it gets hung up and torn off in bedding, upholstery, or shaggy carpeting or grows back into the foot and produces a nasty abscess. At such times both owner and cat become immediately and painfully aware of abnormal nail anatomy. If you own a feline senior citizen, be sure to check all of its claws regularly and pay particular attention to the "thumb" claws, which receive little wear. If your cat can't maintain its nails, either you or your veterinarian may have to lend a helping hand.

TONGUE-TIED FELINE BEHAVIORS

When we think of the cat's mouth, the structure that immediately pops into mind is the tongue. Surely memories of rough laps across hand, cheek, or nose number among people's most pleasurable and frightening feline-related experiences. I count Maggie's reassuring laps in response to her routine brushing among my finest rewards; nor can I forget the icy chill that swept over me when her tongue glazed my bare leg as I inched through the darkened bedroom, convinced she was outdoors.

We already noted the survival benefits conferred by the short, easily groomed coat, and the tongue performs the necessary cleaning admirably. The backward or posteriorly angled barbs not only remove loose hairs but also fleas and other parasites and debris. However, while we certainly can't deny the value or efficiency of this organ, we can't ignore its contribution to that aforementioned feline nemesis: hairballs.

How in the world did the barbed tongue survive when it's perfectly obvious that a smoother tongue would remove less hair and therefore lessen the chance of hairball formation? First, we must bear in mind that structures that persist do so because they are beneficial given the existing circumstances. In the case of the wild feline the need for an effective means to remove dead hair and debris from its coat is pitted against any harm that hair and debris may cause when swallowed and passed through the digestive tract. The obvious solution would be either to decrease the amount of debris swallowed or to ensure the safe removal of that hair and debris from the system. To achieve the former would require major anatomical changes in the tongue's surface and the creation of new behavioral displays that would accomplish the critical grooming some other way. While it is theoretically possible given sufficient time and genetic alterations, Mother Nature provided a much more readily available and elegant solution. Like many carnivores (meat eaters), cats prey primarily on herbivores (plant eaters). In addition to providing a well-balanced diet, many people believe that the plant matter that fills the rodent stomach and intestine serves as a natural lubricant and laxative. Others postulate that the plant matter may contain enzymes that enhance the cat's ability to digest complex proteins such as those found in hair. Either way, the diet plus the short coat would give the wild cat two advantages not available to Figaro the Persian consuming his canned cat food diet.

Feline Bodyworks: The Physical Cat

We can also appreciate how the tongue's roughness serves as a powerful stimulus to newly born kittens, causing them to let out irritated gasps and yowls that clear their lungs of fluid and trigger normal respiration. As the kittens mature, they associate the tongue with grooming and positive interaction with members of their own species. Perhaps this explains why cats also use grooming as a form of displacement behavior; when cats have negative interactions with other cats, some may commence grooming themselves at what appears to be a most inappropriate time. Owners who've surprised their cats napping on the dining room table or grazing on the kitchen counter when "they know better" often notice this identical behavior. The cat ceases one behavior, then begins grooming as though it had been involved in this task for hours. In both the feline/feline and human/feline interactions, the grooming gesture may serve as a behavior the cat uses to signal that all is well, the confrontation is over. Compare this to how people will often treat themselves to a cup of coffee, a snack, cigarette, or drink after a difficult interaction with another person.

As we shall see in our discussion of territoriality, such body-language displays offer easily recognized cues in the wild and prevent strife. However, just as a person's pleasantly symbolic cup of coffee or drink may evolve into an obsession if the initiating stress continually occurs, so displacement grooming can become a serious and frustrating behavioral and medical problem.

MORE THAN MOLARS

The second oral structure of particular behavioral significance is the teeth. Obviously any predator's survival depends on healthy teeth, both to capture prey and to consume it. However, more than one toothless domestic cat has lived a full and happy life eating a balanced diet of canned food.

Whereas the fangs or canine teeth assume the primary role during the capture and killing of prey, the molars directly or indirectly come into play in two other common feline behaviors not associated with

THE BODY LANGUAGE AND EMOTION OF CATS

eating: scent marking and mating/maternal displays. We've already noted how cats use head and cheek rubbing to mark treasured objects and loved ones and how part of the display involves pulling back the lips to expose scent glands in the corners of the mouth. In the process of rubbing and marking, then, the upper molars in particular are apt to be exposed, especially when the cat marks hard or rough surfaces that further hold the lips back and expose the teeth.

In domestic cats it's not uncommon for these back upper molars to become covered with a thick layer of tartar. Although little documentation regarding the cause of this phenomenon is available, logic offers two possible explanations. First, domestic-cat diets may not provide the tooth-cleansing properties of those consumed in the wild. Second, the tartar may be laid down as a protective response to the chronic irritation created by the marking behavior. To me, the latter seems more likely, since it's not uncommon for the upper molars of domestic cats to accumulate tartar even though all the other teeth remain perfectly clean; moreover, this phenomenon occurs in cats fed dry, canned, or semimoist diets. If the tartar were related to the composition of the commercial food, all the teeth would be affected; if it were related to the form of the food, we'd expect that cats fed dry food would not have the problem. Because only the molars are affected and because cats fed hard food do accumulate tartar on these teeth, chances are that diet isn't a major contributing factor.

While the accumulation of tartar in and of itself may perform a passive if not a positive function, it becomes problematic when debris becomes trapped beneath it. The normal population of mouth bacteria then zero in on the debris, resulting in infections that often involve one or more of the molar's roots. Surprisingly, the first indication owners may have that a tooth problem exists takes the form of an apparent *eye* problem. Because the multiple roots of these teeth extend into the area just below the eyes, pus buildup around the roots applies pressure on the lower ocular structures and causes swelling. Infections at this stage usually require anesthesia and extraction of the tooth, a relatively simple procedure usually followed by a speedy and uneventful recovery.

Healthy teeth are also necessary for normal maternal displays such as retrieving or transporting young. Obviously if the queen experiences pain when she grasps her young with her teeth, she's not going to be anxious to perform this often-critical survival function for her kittens.

Similarly sound teeth play a critical role in mating displays. Oftentimes when people own a cheerful, friendly male, particularly one with exceptional physical as well as behavioral attributes, they use the cat for breeding purposes. If the owners observe the cat's liaisons carefully, they may sense the male's growing unwillingness to perform. Although many factors, such as arthritis, could contribute to his reluctance, the wise owner will look at the cat's teeth. Earlier, we discussed how the male grasps the loose skin on the female's neck during mating: If infection or tartar accumulation make this uncomfortable or painful, the cat will quickly lose interest in breeding.

FELINE COMINGS AND GOINGS

Perhaps no feline anatomical feature complicates owners' lives more dramatically than the urogenital system. Screaming tomcats spraying foul-smelling urine or engaging in vicious fights; females in heat emitting mind- and soul-wrenching yowls; urinary tract infections that turn the whole house into a litter-box assault cat owners everyday.

Does feline urogenital anatomy and physiology differ all that much from that of other mammals? Yes and no. As we shall see time and time again in our study of feline body language, although certain physiological structures may look just like those of other species, the cat often uses or expresses its physiology in a dramatically different way.

For example, feline and canine female reproductive tracts compare more than they contrast; however, the breeding cycles and displays of the two species are like night and day. In general, bitches cycle about every six months, their heat lasts about three weeks, and ovulation and

THE BODY LANGUAGE AND EMOTION OF CATS

termination of the cycle don't depend on mating. Most likely because of their closer association with humans and selected breeding for other characteristics, bitches can and do come into heat any time during the year rather than during those optimal mating seasons preferred by their wild ancestors.

Cats, on the other hand, display reflex ovulation and seasonal polyestrus cycles. Ovulation results from copulation or sexual stimulation; if mating occurs, the cycle lasts four to six days and doesn't recur until the next breeding season. If the female doesn't mate, her cycles may last up to ten days and recur every two to three weeks through the season. In the wild, the breeding season and its duration reflect the ideal times and conditions for successfully carrying and raising young. And in fact, many cat owners can tell that spring has arrived as much by the sounds of a rousing cat fight and the stench of tomcat urine as by the robin's cheerful tune or the intoxicating scent of daffodils. As with other aspects of its behavior, the cat's long association with people has altered its reproductive cycle to the point that in some areas a second breeding season occurs during what would not be a climatically desirable time for a wild cat. In my own region, the major breeding season extends from late February through April, guaranteeing that the resultant crop of kittens has the whole summer and fall to grow. However, we also experience another shorter season in late fall. From a survival point of view, there seems no logical reason for a queen to give birth during the depths of the New England winter, so, does this merely reflect a behavioral remnant left over from those ancient feline ancestors who lived in less hostile environments? Or is this a new pattern that evolved when the cat's close association with people provided protection from the natural environment, making such biological safeguards unnecessary? For example, the evolution of a life-style that included spending the cold months in a snug barn with a dependable rodent population might offer winter-born kittens different, but comparable, advantages. As usual, the cat provides us with a paradox, both extremes of which may be truth.

Feline Bodyworks: The Physical Cat

While the feminine urogenital tract is anatomically and physiologically unremarkable, the male system does offer some important variations. Most noticeably, the feline penis points toward the rear when the cat relaxes, but forward when the cat is aroused. Moreover, barbed projections protrude from the intact tom's penis, whereas that of a sexually immature or castrated male appears relatively smooth. Recall how longhaired toms may accumulate a ring of hair around the penis that prevents breeding. In this case, the barbs on the penis serve similarly to those on the tongue, effectively trapping loose hairs. Some researchers feel that these barbs contribute much to the unique orgasmic nature of feline copulation. Paradoxically others see the intensity of the queen's reaction not resulting from this stimulation, but rather from the pain produced when the barbed penis is withdrawn.

A second anatomical idiosyncrasy of the male feline urogenital system has to do with the location of the urinary bladder, which lies further forward in the body than that of other species. This means that the urethra—the muscular tube that carries urine from the bladder to the end of the penis—must be proportionately longer. Furthermore, the urethra becomes progressively narrower in diameter as it travels from the bladder through the penis.

If we consider these three anatomical variations:
- The posterior alignment of the relaxed penis
- The more anterior urinary bladder and associated longer urethra
- The diminishing urethral diameter as it progresses from the bladder through the penis

we can see how male-cat anatomy contributes to that perennial cat and cat owners' nightmare: urinary obstruction. The viruses that cause most urinary tract infections produce mucousy or gritty debris, which is then expelled when the cat urinates. To understand why females are much less likely to block than males, imagine emptying a sand and water mixture from a balloon with a short, wide neck versus one with a long, gradually narrowing neck that also curves. Which system will flow more freely? Which will easily clog? Therefore, even though both male and female cats do suffer from urinary tract infections, basic anatomical differences make such infections potentially more hazardous for males. This occurs because the life-threatening changes result from the accumulation of toxic waste products secondary to the obstruction, not in response to the viral infection itself.

MOUNTAINS OUT OF MOLEHILLS

While the fluid grace of the moving cat would seem to require all sorts of unique muscles, in fact the feline musculoskeletal system bears a striking resemblance to those of other four-legged mammals. However, once again we discover that it's not the cat's anatomy that is unique so much as how it uses that anatomy to create unique body-language displays. We already noted how the cat's hairs are attached to the skin in groups and how each group has muscles that elevate the hairs to increase the coat's insulating effect. These muscles, called the *erecto pili* muscles, are connected to all hairs except the tactile hairs, or *vibrissae*, and they are most highly developed along the spine and tail.

In addition to providing insulation, the elevated hairs serve a critical defensive function. Although we're all familiar with the raised hackles a frightened dog displays, that reaction pales beside the dramatic increase in size the cat can feign when it contracts the *erecto pili* muscles and elevates its hair. The rapidity with which this change occurs can unnerve, disorient, and even drive off a larger opponent. While the anatomical features of the canine and feline reactions are quite similar, the feline display tends to be much more startling.

Sometimes this display so frightens people that they forget that's what it is—only a display the cat uses to make itself appear larger in hopes of frightening off a potential assailant; it's not a display the cat uses to provoke an attack. Nor does the display mean the cat wants to fight, or even that it intends to; it's simply the cat's way of signaling its own fear. Whether or not that fear results in a fight depends on whether the perceived threatener heeds or ignores the warning. The unknowledgeable person who further antagonizes a cat making such a display, angrily thinking, "The cat started it!" fails to recognize that the cat thinks that the *person* precipitated the negative encounter.

THE RESONANT FELINE

If the image of the puffed-up Halloween cat reminds us of the ailurophobic medieval Europeans, that of the contentedly purring feline quickly aligns us with the ailurophilic ancient Egyptians. Who can deny the mesmerizing effect of those rhythmic pulsations, particularly when we're feeling a little down? Consequently it comes as no surprise that little lore associates purring with evil or other negative factors, if for no other reason than ailurophobes are unlikely to find themselves in the company of a purring cat. To be sure, we all know of cats who possess the uncanny ability to identify non-cat people and make their lives miserable with incessant shows of feline devotion and affection, including purrs. However, while these people may distinctly dislike cats, even they often admit that the purr is the animal's one redeeming feature.

It's quite possible that much of the purr's appeal comes from the fact that the anatomy and physiology behind it have been shrouded in mystery for centuries. Even now, with sophisticated equipment and techniques bringing us closer to unraveling the mystery of the purr, we still don't understand it precisely. Consequently, purring simply reinforces our image of the cat as a loving, but mysterious, creature.

Every decade seems to produce a definitive new theory regarding the anatomy and physiology of the purr; however, all more or less agree that the purr involves the hyoid apparatus. The hyoid apparatus developed when early fish evolved into air-breathing forms; the bony structure once necessary to support gills became modified into one that supported the tongue and larynx (voice box) and protected the trachea or windpipe from foreign debris. Early studies indicated that wild felines possessed subtle hyoid apparatus variations that, in turn, led to different vocalization patterns. That of the big cats is more flexible, allowing these cats to roar easily; however, they only purr while exhaling.

Smaller cats have a less flexible, ossified, or bony, apparatus which limits their ability to roar but enables them to purr while inhaling or exhaling.

Although most purr theories agree that the hyoid apparatus serves as a sound transmitter, they disagree about the actual source of the transmitted sound. Until recently the most prevalent theories held that the purr occurred when blood flow in major chest vessels increased. The problem with these blood turbulence theories centers around the established and long-recognized (since 1936) fact that the purr's innervation originates in the brain, suggesting a higher-level control mechanism than that of a simple knee-jerk reflex, for example. If such a mechanism does control purring, then using it to trigger alteration of flow in major vessels (including, according to some theories, the supreme vessel, the aorta) to signal something as physiologically insignificant as pleasure seems both excessive and inefficient. To alter the blood flow (thereby disrupting the flow of vital nutrients and oxygen and the removal of waste products) to signal pleasure is comparable to rhythmically slamming on your car brakes rather than sounding the horn to attract another motorist's attention. While it can be done, the resultant wear and tear on your car far exceeds that which would result if you used the horn. So while blood turbulence theories offered physiologically possible explanations for the purr, they seemed extreme when viewed in terms of the results.

In 1985, Drs. Lea Stogdale and John Daleck compiled data supporting the existence of a more passive mechanism, one involving airflow. When our ancient fish ancestors made the switch from water to air, they needed some way to keep solids and liquids from getting into their lungs; and these land dwellers also needed a different form of communication. In an exquisite display of evolutionary economy and harmony, the gills of the fish gave way to the protective respiratory structures that further evolved into the larynx and vocal apparatus. While the resultant gear appears to have many parts (and naggingly similar, but unpronounceable Latin names), it functions in a beautifully simple way. A cartilaginous flap, the epiglottis, opens to permit the

Feline Bodyworks: The Physical Cat

flow of air into and out of the larynx by way of the glottis, which also opens and closes in response to air pressure. Immediately behind the glottis lie folds of soft tissue, the vocal cords, which vibrate rather than open and close in response to changes in air pressure. At the end of the acoustic setup we find the trachea, a hollow, air-filled tube that carries air to and from the lungs.

Take a deep breath, then exhale slowly. Notice how inhaling creates air pressure moving toward your lungs, whereas exhaling produces airflow and pressure in an outward direction. Because all three anatomical structures respond to changes in air pressure, we can see how theories that locate the purr in this area allow the purr to exist within the mechanism of normal breathing and during inhalation and exhalation. They also eliminate the inconsistencies present in the major blood-vessel-vibration explanation.

But how does the cat actually produce its purr? Whenever the glottis narrows, air pressure increases in the larynx and on the vocal cords. The motion of the latter causes the air to vibrate and resonate throughout the entire respiratory system. Imagine bursting into an empty room through two swinging doors: The doors continue to flap behind you and the sound reverberates throughout the room. If empty rooms exist on both sides of the swinging doors (such as the spaces on either side of the vocal cords), regardless of which way you pass through the doors, noise will occur. If you could do this fast enough, the noise would sound like a continuous flapping. Similarly, when the vocal cords begin vibrating and relieving the pressure, the glottis opens and permits airflow either into or out of the system. Because the entire sequence occurs in only thirty to forty milliseconds (thirty to forty *thousandths* of a second), we hear a continuous sound.

Although this theory eliminates many vaguenesses of past explanations, it doesn't tell us how the nerves trigger these changes or why such changes occur in some species but not in others. While the prevalence of purring during intraspecies acts such as nursing and mating, and interspecies interactions such as petting and grooming, tempt us to view purring as a most positive form of communication, we stumble over our awareness that many cats also purr when they're terminally ill and right up to the moment of death. However, while this type of purring might baffle the objective observer, it does make sense if we view purring as the most *intimate* form of communication. Like the Egyptians and witches of the past and present, all cat owners have observed their pets seeing the unseen movement and hearing the sound beyond human comprehension. Consequently, we might conclude that the purr, too, reflects interaction on levels currently beyond our comprehension.

While our objective minds might cringe at such ideas, our own experiences often keep such possible, if inexplicable, subjective explanations alive. For example, whenever I attempt to unravel the enigmatic purring behavior, I'm reminded of my father and a former employer of mine. Whenever Dad performs a task he finds enjoyable or one that requires concentration, he's apt to hum some meaningless (to me) sequence in his deep bass, which sounds not unlike a purr. If I ask him what he's humming, more often than not he responds, "Who's humming?" My former employer gave his staff audible warning that trouble was brewing because he always whistled the same ditty off-key when stressed; he, too, was usually unaware of this behavior. In addition to these human "purrs" related to concentration and stress, some also experience the more joyful version we usually associate with the cat. The latter is the form I catch myself using: If I've received good news or shared a positive intimate interaction with a loved one, I'm bound to catch myself humming, usually nothing in particular. So while we may tend to think of the purr as a uniquely feline behavior, it may be that it appeals to us because it's so similar to our own.

THE CENTER OF THE FELINE ANATOMICAL UNIVERSE

We're going to conclude this chapter with a brief discussion of some characteristics of the feline central nervous system (brain and spinal cord) that contribute to the cat's unique body-language displays. What

Feline Bodyworks: The Physical Cat

about that characteristically graceful feline movement—aren't there special muscles and/or bones that enable cats to move in such distinctive ways? Actually, much of the smooth and agile movement we associate with cats result more from variations in its central nervous system (CNS) than any musculoskeletal variations. Unfortunately there are little data available comparing the nervous functions of different species in any great detail. It is possible, however, to match dog and cat bones and muscles one for one, and few would argue that members of these two species do move differently. For example, although both cats and dogs are capable of jumping from a seated position and use essentially the same musculoskeletal apparatus to accomplish this feat, the cat's leap certainly appears smoother, more graceful, and more effortless when compared with a beagle's, and most certainly a bulldog's.

The way in which much of the central nervous system actually functions still evades scientists, but several studies do provide some intriguing insights. When considering the cat's incredible grace and agility, it's interesting to speculate on the role the spinal cord plays in the process. In order to do so, we need to compare the vertebral columns vis-à-vis the length of the spinal cords in three mammalian species— human, canine, and feline. First, notice the numbers and kinds of vertebrae that make up the vertebral columns of each species:

SPECIES	NUMBER OF VERTEBRAE				
	CERVICAL	THORACIC	LUMBAR	SACRAL	COCCYGEAL
Human	7	12	5	5	4
Canine	7	13	7	3	6–23
Feline	7	13	7	3	6–22

Most mammalian vertebral columns contain seven cervical or neck vertebrae followed by twelve to thirteen thoracic vertebrae, to which the ribs attach to form the chest cavity. Below these in humans and behind them in the dog and cat lie the lumbar vertebrae, which form that part of the spinal column between the chest and the pelvis; the sacral vertebrae are fused and incorporated into the pelvis itself. At the end of the vertebral column sit the coccygeal vertebrae, sometimes called the tail bones. Depending on the individual and breed, the first coccygeal vertebra may also be fused to the sacrum. In dogs and cats the number of coccygeal vertebrae determines the length of the tail. Breeds such as the Manx cat and pug dog have fewer coccygeal vertebrae than Siamese cats and Irish setters, with their characteristically sweeping tails.

The all-important spinal cord lies sheathed within the protective bony canal created by the vertebrae. Think of the spinal cord as the main pathway linking the brain and the countless peripheral nerves controlling the body. If the brain were a bustling metropolitan area, the spinal cord would be the one well-designed, well-maintained thruway leading in and out of the city; and the peripheral nerves would be the side streets to suburban homes. Extending this analogy, the closer one lives to the thruway, the more efficiently one can travel between one's home and the metropolitan area.

The spinal cord transmits information to and from the brain. Therefore it would seem that the length of the cord relative to the vertebral column would give us some idea of the relative innervating efficiencies of various species. If the muscles and skeletal structures were identical in two species but the spinal cord of one was relatively longer and therefore provided easier access for nerve impulses, perhaps such animals could be capable of finer or more complex movements. As it turns out, the human spinal cord terminates at the level of the first or second lumbar vertebra, the dog's between the sixth and seventh lumbar vertebra, and the cat's between the seventh lumbar and the third sacral vertebra. Actually, various researchers describe the feline cord terminating as far down the spinal column as the third coccygeal vertebra. Such inconsistencies in data arise because the cord migrates posteriorly as the cat matures; the cord of a six-month-old cat may terminate more anteriorly than that of the same animal a year later. Unfortunately, because consistent data regarding age do not accompany the research, we can't pinpoint the location exactly. However, whether the feline cord terminates at the third sacral or the third coccygeal vertebra, the fact remains that it extends beyond that of the dog and most certainly our own.

Imagine this typical scene in my yard. A butterfly pauses to rest on the edge of the picnic table, barely waving its colorful wings. Attracted by the motion, Maggie freezes at her perch on the back of the couch in the living room. Then she races soundlessly to the kitchen door, leaps, and sinks her claws into the door's paned windowframe and lashes her tail rapidly—her way of saying, "Let me out." As I open the door, she leaps, spins midair, and somehow gets around me and out the door in what appears to be one graceful motion. Then she stalks the butterfly, at times holding her body so still that it would seem to be painful. Suddenly one paw moves forward, and another comes off the ground so quickly that I feel the motion rather than actually see it. Then she leaps, but this time it's a long, shallow leap, more like flying. Her front paws flare and her claws arc and miss. She lands, twists, and dances on her hind feet, takes a few more playful swipes, then settles gracefully on her haunches and licks her shoulder thoughtfully as the butterfly glides away. Much of this captivating yet familiar feline display relies on the almost uncanny way the cat can manipulate its spine, hindquarters, and tail.

While most of the differences between the feline and other mammalian brains result from differences in sensory apparatus, which we will be discussing in chapter 3, some make their presence known in rather obscure ways. Like footprints, which can tell us *something* about the shoe and its wearer but certainly not everything, the cat's reaction to some drugs that affect the brain indicates that the feline brain differs markedly from that of a little dog, but it doesn't tell us exactly how it differs. For example, veterinarians regularly use barbiturates as general anesthetics. These drugs depress the central nervous system, and although we know little about the exact site and mechanism of action, we do know that they affect parts of the brain. Cats given comparable doses of barbiturates tend to become more depressed and sleep longer than other species. Furthermore repeated doses of these drugs tend to produce a reverse tolerance: With each successive dose, the cat experiences a deeper level of anesthesia and takes longer to wake up.

Another unique drug response that supports the view of the unusual nature of the feline brain involves its response to the drug morphine. Morphine is believed to act on the same receptors as the body's own natural pain-relieving substances, called endorphins. Studies indicate that those given morphine experience diminished pain partly because the drug diminishes the *response* to pain—the anxiety, fear, and panic—rather than the painful stimulus itself. In other words, morphine is a mind-over-matter drug, creating a mental condition that permits a more favorable response toward and acceptance of a painful physical condition.

Although scientists have worked out the pharmacology of morphine over many years of careful study of its effects on different mammalian species, the theory breaks down completely when it comes to *Felis domestica*. Given comparable doses, not only doesn't the cat experience the characteristic euphoria exhibited by other species, it experiences the opposite extreme: intense excitement and even mania.

Why do cats exhibit these unique responses to drugs such as barbiturates and morphine? That very good question unfortunately leads to other very good questions. Perhaps the cat's awareness of its surroundings is such that it can override its awareness of pain more readily than it can adapt to substances that interfere with its perceptions. Evidence that some people believe that cats are capable of experiencing voluntary altered states may be found in Indian legends concerning the origin of hatha-yoga. According to one of these a young Indian prince, frustrated by his inability to meditate, retreated to the forest, where he met a cat who taught him the basic relaxation postures of what was to become hatha-yoga. While such tales may or may not be true, the fact remains that the yoga masters are certainly capable of producing mental states which enable them to perform often extraordinary mind-over-matter feats. Common sense suggests to even the nonmedically oriented that individuals capable of these deeply relaxed states might respond much more dramatically to barbiturates and other depressants simply because they're more relaxed (that is, physiologically slowed down or depressed) to begin with. Similarly, if cats can naturally assume the trancelike state of a yogi, then the presence of a barbiturate would actually intensify this normal feline response rather than create it. Compare attempting to lull a wide-awake child to sleep with soft music or bedtime stories versus one who's already nodding off; which one requires more effort to calm to the point where sleep occurs?

Likewise, if CNS depressants may precipitate exaggerated responses in a normally relaxed individual, it's possible that drugs mimicking other CNS functions might exaggerate those functions, too. Perhaps the cat's normal "extrasensory" perception, which some researchers have investigated (see chapter 6), becomes unmanageable when exaggerated by drugs such as morphine, which depresses the brain but stimulates the spinal cord. Unfortunately, although some people react to morphine in the catlike manner rather than with the much more common doglike human response, no correlation has been made between that exaggerated response and any human extrasensory abilities. If such a correlation did exist, if these people and cats do share a common brain pattern or function evidenced by their paradoxical response to morphine, then the idea of witches and their cats being on the same "wavelength" or "beam" might be more truth than fiction.

One thing is for certain, the unhuman and undoglike response of cats to many common and not so common drugs makes it crucial that nothing other than those products labeled specifically for cats ever be given to your pet without veterinary approval. If your cat should be deliberately or inadvertently exposed to substances, legal or illegal, known to affect the central nervous system, contact your veterinarian immediately and do not withhold information. Nothing is more frustrating to a veterinarian and life threatening to the cat than owners afraid to admit the true identity of the "uppers" or "downers" or "weed"

THE BODY LANGUAGE AND EMOTION OF CATS

consumed by their pet. Only with this knowledge can the clinician implement the specific antidote, if one exists. Regardless of the practitioner's personal feelings concerning the events that led to the cat's condition, the majority see their role as helping the animal, not passing judgment on the owner. Even if you do have to endure a bit of ranting and raging by 'fessing up, surely it's worth your cat's life to do so.

While the feline anatomical and physiological variations discussed in this chapter do help set the stage for our study of feline body-language displays and our responses to them, to understand this intricate creature fully, we need to delve into the most fascinating anatomical and physiological system of them all: the sensory system. In the next chapter we'll explore the cat's senses. Do cats see, smell, hear, taste, and feel differently from us? How can we possibly hope to understand what's going on in that furry little head if we're not capable of seeing and hearing things the same way?

3

FELINE SENSATIONS: READING THE MIND OF THE DEMON-GOD

*T*HIS cat's nuts," Lisa Mc-Dermott confides to her best friend, as she gestures toward her gray, white-footed domestic shorthair, Spats. "Sometimes he sits and stares at the bookcase in the living room for ages, even though I *know* there's nothing there. Other times the house is perfectly quiet and he flies to the window like he's possessed. Spooky!"

Several weeks later when Spats turns up his nose at yet another variety of food after sniffing it with disdain and giving it the tiniest lick, Lisa throws up her hands in disgust.

"I've had it. You didn't even taste it and it cost me an arm and a leg. You're either crazy or stupid, Spats, and something's got to change before *I* go nuts!"

Much of the mystery we associate with our cats derives from the anatomy and physiology of their senses. Although how they see, hear, smell, taste, or feel may not differ all that much from other mammals, their senses do function quite unlike our own. If we pay attention to how the cat perceives its world, we can go a long way toward effecting the kinds of changes Lisa thinks she needs to make to create a more workable relationship with her cat. This doesn't mean we must learn to see, hear, smell, taste, or feel like our cats in order to relate to them properly. However, an appreciation of how their senses compare to our own reveals a lot about previously incomprehensible or even unacceptable behaviors. As we noted earlier, the differences that distinguish cats and their relationships with humans on all levels contain some surprises and paradoxes, and nowhere do we see more of them than on the sensory level. Just how do cats use their anatomy and physiology to interpret or respond to their world? Let's briefly examine each of the feline senses to discover some of the many answers to that question.

THE DOUBLE-DUTY FELINE OLFACTORY SYSTEM

The feline olfactory apparatus functions much like that of the dog. Like all mammals except New World monkeys and man, cats use two systems to collect scent data, which are then sent to the brain for analysis and interpretation. One system links the vomeronasal organ (whose duct opens at a small bump just behind the upper incisor teeth) to special olfactory areas in the brain. A second system connects receptors and nerve fibers in the nose to the olfactory region of the brain, but not the same area that oversees the sensations and responses of the vomeronasal organ. Although both systems respond to scents, the cat uses each for specific purposes.

When cats wish to utilize the vomeronasal system, they lift the upper lip to expose the duct opening and simultaneously block the nasal passages and the other olfactory apparatus. This display, called the flehmen reaction, accounts for the rather vague, entranced expression that comes across a cat's face when exposed to certain scents. While most researchers associate the vomeronasal organ and flehmen reaction with the male's perception of the specific scent hormones (pheromones) exuded by females in heat, both sexes utilize this system when examining the urine and scent marks of other animals.

Significantly, the feline vomeronasal olfactory system also responds to at least one pseudosexual odor. Most people who observe a catnip-responsive cat describe the behavior as "euphoric" or "drunk." However, anyone who's observed females in heat when they smell tomcat

Feline Sensations: Reading the Mind of the Demon-God

urine, or their behavior immediately after being bred, can't help but notice the similarities between these displays and the catnip response. Research indicates that catnip contains a hallucinogenic compound, neptalactone, which affects cats the same way their natural sex pheromones do. Although domestic cats exposed to catnip eventually develop olfactory fatigue (that is, after a period of time the brain no longer responds to the odor), some large wild felines can actually become addicted to these plant substances. Most fascinating of all, biologists have recently identified a dominant gene that enables the cat to locate the plant in its environment. Presumably a similar gene exists in the 50 to 75 percent of the domestic cat population that reacts to catnip.

While such data could seemingly support a definition of the cat as a natural drug addict, that really doesn't make much sense. In terms of species survival, solitary species must develop a reliable system whereby receptive females and males can locate each other. Because clear visual and auditory signals could easily attract predators—with the signaler winding up as somebody's lunch rather than somebody's lover—cats have developed subtler olfactory cues. Studies of domestic cats reveal that while they react strongly to the presence of the sex pheromones in other cats' urine, they don't get particularly excited over that of other species, even when a member of that species is sexually active. Consequently cats responding to neptalactone aren't responding indiscriminately at all, but rather quite discriminately to a precise botanical mimic of a natural physiological compound.

An obvious question is why do these plants exude this particular odor? We know that some plants use such scent mimicry to lure specific insects to them as a means of disseminating pollen. It could be that neptalactone-containing plants use their scent to lure wild felines to function as furry, earthbound bumblebees. More interestingly, we can't overlook the possibility that these plants could also function as a natural form of birth control and selection for wild felines. Those animals who can't find mates during the breeding season can turn to this natural release mechanism; and those who prefer the plant form

THE BODY LANGUAGE AND EMOTION OF CATS

and become addicted to it soon lose interest in mating and survival.

Although the cat's sensory system can outsniff a human's any day of the week, domestic cats don't rely on either aspect of their olfactory mechanism that much for their survival. Cats who have lost their sense of smell still copulate normally, thus indicating that the vomeronasal system isn't critical to their reproductive ability. Moreover, cats rarely follow scent trails, tending to prefer the sound of prey (or the can opener) to the smell of the food. Only when the cat wants to locate or investigate stationary, silent objects (such as a small, immobile bit of food or a toy) does the second half of the olfactory system, collecting data via the nose, kick into full gear.

By now we know that treating any feline system as insignificant or totally predictable invariably invites the exception to the rule. While wild cats depend minimally on their noses to locate and catch prey, average house cats, whose precaptured food comes in a bag or can and more or less magically appears in a bowl, depend on them a great deal. Most veterinarians and owners confront the olfactory nemesis in its most common form: the cat with an upper-respiratory infection that refuses to eat. Because the cat can't smell its immobile food, it won't eat; because it won't eat, its ability to fight off the infection decreases and its condition worsens. When Spats comes down with a cold, sometimes all Lisa needs to do to keep him eating is to provide smellier fare—canned food instead of crunchies, Mackerel Garlic Goop instead of Creamed Chicken Surprise. Other times she must stick the food in Spats's mouth. Although cats with upper-respiratory infections and other ailments may lie next to the dish of food as though it were nonexistent, if you dab a bit on your finger and smoosh it up against the roof of the cat's mouth, many a cat will perk up and begin licking and swallowing. Once this occurs, many cats will then act as though they suddenly remembered something and start eating on their own.

How on earth could any animal forget something as critical as eating? When we study predation, we'll discover that it's a four-step process, each step of which requires a stronger external stimulus than the

Feline Sensations: Reading the Mind of the Demon-God

one before. Because eating represents the final phase of the predatory display, it results from a full range of increasingly stronger sensory stimuli. When these stimuli don't appear in the proper forms and/or amounts, the cat won't eat.

Before we grieve for the cat who doesn't eat because it can't smell, we must consider the other half of the story. Research indicates that in the initial stages of an illness, not eating actually facilitates the immune response and helps the cat fight off infection. But what if the cat clicks off its appetite center, literally forgetting to eat until reminded? In wild felines such a reminder probably takes the form of a scurrying mouse that teases the more sensitive visual and hearing senses into alertness; and these in turn initiate the predatory sequence. Cut off from this natural source of stimuli, domestic cats must often rely on their owners to supply sufficient stimuli in the form of smellier foods, force-feeding, or even medication to trigger the eating response.

Because the wild hunting cat collects a much broader range of sensations to stimulate its appetite, smell isn't nearly so important to it as it is to the house cat, whose mouse-shaped crunchies and creamy tuna pâté neither move nor make noise. Moreover we can't overlook the possibility that the discrepancies between wild and domestic feline olfactory anatomy and physiology may become more pronounced as we continue selective-breeding cats to satisfy human criteria.

For example, animals who obviously collect scents from moist body openings or secretions often strike people as behaving repulsively. We need only observe some people's reactions to cats sniffing and licking their own or another cat's "private parts" to realize why cats who don't do that might appeal to these people as preferable pets. Consequently, people may have consciously or subconsciously bred nonsniffers over the years; and as long as those people cared for their cats, this selective dulling of olfactory sense caused no obvious problems. Therefore it's possible that over the ages, the vomeronasal system in particular played less and less of a role in what was and is considered normal cat behavior. That such is entirely possible was made dramatically clear when

THE BODY LANGUAGE AND EMOTION OF CATS

Lisa rushed Spats to the veterinary clinic because he was having "a weird kind of seizure, acting stupid with his mouth hanging open and his eyes shut." With our knowledge, we immediately recognize this as the typical and quite normal flehmen response; yet Spats and Lisa have lived together six years, and this is the first time Lisa noticed the behavior.

Paradoxically, the modern cat's olfactory system may rely more on the collection of scent data through the nose than its ancestors, primarily because of their different eating habits. Although smell plays a minimal role in the predatory sequence, it does become important in the case of stationary food. If people associate love or affection with an eager eater, cats exhibiting the latter behavior would once again make more appealing pets. In such a way an olfactory orientation that might distract an animal in the wild may serve it well in a domestic environment.

PASSING THE FELINE TASTE TEST

During an animal nutrition course, one of my professors addressed the subject of finicky feline eaters with a memorable observation: Any animal that routinely eats raw mice isn't by nature a finicky eater. A wild creature's taste, like all its senses, provides survival advantages for the individual and the species. As such, it need only distinguish safe from unsafe, beneficial from nonbeneficial. Moreover, such a system must function simply and effectively.

Recall the children's game in which a whispered message travels from child to child: The more complex the message, the less likely it will emerge in its original form at the end of the line. The same holds true for the young cat, who must learn a great deal from its mother in a relatively short time. Once on its own, learning to detect safe foods by trial and error makes every meal a potential life-or-death situation. Given the risks inherent in such an approach, the cat naturally found a better way. Because mature cats prefer solitude, they must learn basic survival skills from their mothers during kittenhood. During that time

Feline Sensations: Reading the Mind of the Demon-God

the queen brings them food, then teaches them how to hunt that food. Over the centuries, a system evolved that enables a kitten to establish a sensory data bank against which it compares all potential foods throughout its life.

Let's observe Spats's wild cousin as he evaluates a half sandwich left by a careless camper. Because the object is immobile, the wild cat first sniffs it in an attempt to collect sufficient scent data to identify it as edible or not. In this case, sniffing doesn't provide the confirmatory data, so the cat gives the morsel a few tentative licks, just enough to stimulate its taste buds and provide the minimal data needed to compare this possible food with that presented in the queen's master program. If a match doesn't occur, the cat refuses the meal.

It's a beautiful and elegant system, hardly the work of a devil intent on turning cat fanciers into mealtime slaves. So how in the world did the cat evolve from a survival eater to one with seemingly finicky eating habits? When did the cat stop judging food in terms of safe or unsafe and start reacting in terms of sweet, sour, bitter, salty, fishy, or creamy? Again, we have only ourselves to blame or congratulate for this modern adaptation of an ancient sensory mechanism.

Although palatability—discriminating foods strictly on the basis of taste—plays an insignificant role in feline survival in the wild, humans have elevated it to an art form. Because we project our beliefs onto our pets, and because we and not our pets trot to the grocery store, cat food manufacturers work tirelessly to carve a bigger slice of the $2 billion cat food market with "new and improved" foods that appeal to us first and our cats second.

Figuring out what humans expect in their cat's food proves a more simple task than unraveling the mysteries of something as elusive and idiosyncratic as the feline palate. A legion of about twenty-seven hundred cats in labs all over the country undergo various tests every day as cat food manufacturers struggle to identify which qualities Kitty prefers in her food. The four major cat food conglomerates—Ralston Purina, Heinz (9 Lives), Carnation (Friskies), and Kal Kan—all conduct programs similar to that of Carnation, whose cats perform about 3,000 tests and accept or reject 250,000 cans and 70,000 pounds of dry food every year.

While these facilities have contributed a great deal to what used to be a meager store of information regarding feline nutrition, their research emphasizes palatability as well as nutritional value. What keeps researchers in these test labs awake nights takes the form of a deceptively simple question: What will cats eat that is nutritional, palatable, *and* acceptable to their owners? Mice superbly fulfill the first two criteria, particularly if cooked to destroy disease-causing microorganisms and parasites against which our pampered felines have lost their natural immunity. However, that third criterion keeps many a researcher tossing in his or her sleep—and all those test cats munching and crunching day after day. Not only do owners want their cats to rush to their bowls as if Julia Child had prepared the food herself, they also want their pets to consume foods that appeal to their own human senses.

At the risk of further complicating the taste issue, consider the delicious paradox surrounding the development of feline taste preference. In *The Dynamics of Behavioral Development* (Random House, 1967) Z. Kuo describes an elegant experiment designed to demonstrate whether finicky felines are made or born. Kittens separated from their mothers immediately after birth received specific controlled diets, first in liquid, then semiliquid, then solid forms as they matured. On group ate only a balanced soybean diet, another only mackerel and rice, and the third a wide variety of foods.

When the cats reached six months of age, researchers tested the kittens' preferences for other foods. While it came as no surprise that the kittens fed a variety of foods would eat just about anything, the persistent refusal of those raised on limited diets to eat anything else—even when very hungry—was a surprise. The mackerel/rice-fed kittens would accept canned sardines and red salmon, possibly indicating a highly specific fish-related component in all three diets, but nothing else.

These experiments, while perhaps not conclusive, offer cat owners two valuable and useful insights into feline taste preference:

Feline Sensations: Reading the Mind of the Demon-God

- It develops at an early age.
- It's strongly imprinted and difficult to change.

In fact, taste preference develops in domestic cats during the same span of time the wild queen would use to introduce prey to her young and teach them how to hunt.

When we put these observations together with what we learned about cats with upper-respiratory diseases whose appetites wane because of loss of smell, we can see why sick pets who've always eaten one kind of food can generate some tough feeding problems for their owners. The cat may steadfastly refuse any sort of dietary change, regardless of how critical the change may be to their health. Similarly beef, milk, fish, and eggs, which may be the mainstay of a domestic cat's commercial diet, may precipitate allergic responses in older animals; if the cat has dined strictly on such fare during the crucial developmental period, finding a nonallergic diet may turn into an expensive and time-consuming quest.

These experiments indicate that people who feed cats restricted diets must accept responsibility for creating finicky eaters. The unsuspecting owner who adopts a cat whose previous owner fed it only lobster Newburg makes the cat food marketers' dreams come true. Nothing else may ever please that cat, yet few people are willing to accept this reality, so such owners will eagerly snatch up every new product that comes along. Each new can or bag of food creates its own little melodrama. Spats sniffs; he licks once or twice. Lisa holds her breath: Will he like *this* one? It all depends on what happened during a relatively short span of time years ago.

Or what about those strays who went from mother's milk to mice but whose kind-hearted owners now want to atone for their pet's "miserable" beginning? They, too, end up buying one can of everything on the cat food shelf. Paradoxically, cat owners aware of this feline idiosyncrasy, will also eagerly snatch up every new product in hopes of accustoming their cats to a wide variety of foods and thus avoid future feeding problems. No wonder we spend so much on cat food every year!

The Body Language and Emotion of Cats

When we combine the cat's tendency to rely more heavily on smell when evaluating nonmoving "prey," its habit of chewing its food rather than gulping it like a dog, and its aversion to odors and taste not presented during that critical early orientation period, we can also see why putting medication in cats' food yields mixed results. Because the cat will most likely smell its food, it's more likely to detect any foreign odor. Because it nibbles its food, it may very well taste medications that have passed the smell test. If either smell or taste doesn't register in the "safe" data bank, the cat won't eat the food. Now the owner must contend with a suspicious noneating feline as well as one requiring medication. Because of this, it's much wiser to have your veterinarian show you how to put any oral medications directly into your cat's mouth.

Their beliefs about feline taste allows cat worshipers plenty of opportunities to tithe by spending a disproportionate amount of the family food budget on cat food and to grovel, whine, cajol, and even do little dances to get Spats to eat. These human and feline behaviors in turn infuriate the ailurophobes, who see such displays as the epitome of twisted human values, an overindulgence that only the most despicable animal species would ever tolerate. However, most of us prefer a more balanced approach to feeding, one that permits us to feel we're meeting our pets' nutritional needs without being manipulated or induced to perform mealtime somersaults. By recognizing the role taste plays in the wild animal and the time frame that determines its development, smart owners can avoid or solve the most common dietary problems.

Feline Touch: Winning by a Hair

The cat's tactile hairs, also called whiskers or vibrissae, supply us with another stimulating source of feline fact and fancy. Each of these highly specialized hairs—located above the eyes, on both cheeks and the upper lip, under the chin, and on the back (carpal area) of the front legs—contains highly developed nerves sensitive enough to detect even minor changes in the environment. Researchers believe that the whiskers comprise an elegant sensory system that not only tells the cat what's nearby, but also its own position relative to that object. For example, because cats grab their prey with their front paws, the sensitive carpal hairs can instantly communicate valuable data about the prey's position relative to the pinioning claws. If the claws are imbedded in an object without the carpal hair being stimulated, the cat instantly knows that either the prey is very small or it's being grasped at one end or the other. In the case of larger prey, such as rats, this can be essential information to ensure the kill and also to prevent damage to the cat itself. Not only does a weak or poorly located grip make it more difficult to effect a kill, the prey could escape and/or lash out in defense.

We've all watched cats whisk through openings with barely a hair's breadth of extra space. Because the whiskers on the cat's upper lip tend to be as wide as the cat's body, if they fit through an opening, the rest of the cat probably will too. However, bear in mind that this relationship may not hold true for certain individuals whose "survival" in the show ring might necessitate having an exceptionally narrow head or wide body consistent with certain breed standards rather than the whisker-to-width ratio most consistent with survival in a predatory world. Furthermore, in spite of trends toward human slimness, the fat cat still serves as a positive symbol for some owners; and when such animals' sides outgrow their whiskers, they're apt to find themselves wedged in drainpipes or under couches. Imagine having sensors on the front fenders of your car that tell you whether there's enough room to clear the garage door opening. The system only works if the width of the car's front bumper equals or surpasses that of the back; if Junior opens the rear door, it will snag on the doorframe.

Although owners routinely trim the whiskers of canines belonging to certain breeds for appearance' sake and with little consequence to the dog's welfare, doing so to a cat could affect its ability to orient itself within its environment. Even if an owner wanted to clip a cat's whiskers, most cats hate having them manipulated, particularly in directions counter to the hairs' natural alignment. Although this might not hurt the cat, it does trigger an awareness that resting cats in particular seem to find irritating, possibly because the system normally functions when the cat is fully alert and active. Imagine someone shouting "Look out!" from behind you as you semidoze on a hammock versus when you're playing a fast-paced game of volleyball; the identical stimulus creates two completely different reactions, the former much more bothersome than the latter.

While such a sensory system may add little to the quality of life experienced by Spats, who spends most of his time curled up on Lisa's bed, it may add years to the life expectancy of a nocturnal predator with motion- rather than detail-sensitive vision. In order for a cat to sense something visually, that object must first move. If it doesn't move or make noise, the cat must smell it. Sniffing not only takes time, it makes noise, which may alert the keen-eared prey. Compare these rather cumbersome mechanisms for finding one's way quickly and silently in the dark to one stimulated by the lightest change in pressure. Under those circumstances the touch- or pressure-sensitive system enables the cat to bypass all sorts of threats with the least amount of effort.

RESTRAINING THE TOUCHY FELINE

Although dog books talk a lot about the touch- and pressure-elicited canine defensive response, few cat books talk about it at all. However, we do know that applying pressure to the neck region often elicits a defensive response in young pups and dogs, whereas it usually produces a freeze response in cats, similar to that assumed by kittens being transported by their mothers or by females being bred.

Do these typical responses indicate that cats readily accept restraint? Not at all. And, as we should expect, two quite opposite forms of restraint provide the maximum feline immobility. The first form adheres to the dictum *Less is more.* Generally, the less you restain a cat, the more it will allow you to do; the more you try to restrain it, the more it will struggle and the less you will accomplish. Owners who

Feline Sensations: Reading the Mind of the Demon-God

assume that cats won't tolerate grooming or having medication put into their eyes or ears often ascribe to the questionable logic that, "I know *I* wouldn't like it if I were Pussywillow." Because they assume the cat will find the interaction disagreeable, they further assume the cat will resist and try to run away. Consequently, they grab the cat firmly.

While we can't predict precisely how an individual cat will react to being medicated, we do know that most cats will fight restraint. Unfortunately owners who associate the cat's struggles with the medication or grooming rather than the restraint itself get themselves caught up in a vicious cycle. When the cat resists the medication or grooming, they restrain it even more—which only causes the cat to react more violently. The cat quickly comes to associate the tube of medication or brush with the restraint and bolts at the sight of them. Owners then throw up their hands in disgust, proclaiming, "Pussywillow hates to be medicated (groomed, bathed, petted, etc.)."

While it's very difficult to reverse such a pattern once it's established, it can be done. Simply keep reminding yourself: Less is more. If you begin losing your temper and start aching to squash Pussywillow against the table and stuff the pill down her throat, back off and calm down. Then try again. Sometimes draping a lightweight towel or long, loosely woven scarf around the cat's neck provides sufficient diversion to allow a quick squirt of medication into eye or ear; the soft filmy material will also passively ensnare any stray claws long enough for you to accomplish your mission.

If you feel you must actively restrain your cat, either because it has always behaved unmanageably in the past or because your own fears won't allow you to act confidently without immobilizing it, for heaven's sake do it right. The second, and seemingly paradoxical, rule of feline restraint dictates: *Go all the way.* Maximum restraint tells the cat from the outset that it can't escape, so why even try? To achieve this form of restraint, wrap the cat snuggly in a towel. While thick, loopy materials work well to passively engage claws during minor procedures, they don't wrap easily or securely, and cats will often squirm within such a cocoon, becoming more frightened or angry with every move. Using a more tightly woven wrap removes the option of flight or fight from the cat's repertoire. Anything less will only induce the cat to struggle more; anything more could harm the cat. Because proper restraint is as much an art as a science, have your veterinarian or veterinary technician demonstrate the proper techniques before trying them yourself. Better yet, concentrate on building your own and your cat's self-confidence enough to make restraint unnecessary. You'll be amazed how much your cat will let you do, so long as you don't try to force it.

IN A CAT'S EYE

One day my youngest son, Dan, emerged from his darkened room with a flashlight in hand and a cat under his other arm. "I know how cats can see in the dark," he announced confidently. "They have little lights in their eyes." Whereas the ancient Egyptians saw the cat as a night-stalking portable reservoir of the sun god's great power, Dan envisioned his cat as a fur-covered feline sports car, headlight shields up, high beams penetrating the night.

Both Dan's and the Egyptian's view of ocular physiology may sound fanciful, but they're not that far from the truth. There *is* light in a cat's eye, but it's light reflected from its surroundings. When light reaches the back of the eyeball and strikes the highly reflective tapetum, the excess light energy that doesn't penetrate the light-sensitive receptors of the retina gets "recycled" rather than absorbed by non-functional tissue. In addition to making maximum use of the available light, this bouncing helps distribute the light more evenly across a wider area of the retina.

The proper distribution of the light facilitates nighttime vision. Like all mammals, the feline retina contains two kinds of receptors:

- Color- and detail-sensitive cones
- Motion-sensitive rods

Cones tend to respond individually and require more light stimulation;

rods function in groups and are activated by much lower levels of light energy. Like those of most four-legged mammals, cats' eyes contain more rods than cones. Therefore, although they may be able to distinguish some color and detail, they're much more adept at detecting motion. The reflective tapetum, which not only conserves the low-energy light capable of stimulating the rods but also diffuses it so that as many rods as possible receive stimulation, greatly enhances the cat's motion sensitivity. Compare this with the anatomy and physiology of the much more color- and detail-dependent human visual system. Instead of containing a highly reflective layer to intensify and diffuse light, the human retina possesses a black pigment layer to *prevent* this phenomenon. Because the cones require higher levels of direct energy, any kind of diffusion or reflection interferes with the strong contrast between dark and light necessary to form precisely detailed visual images.

Although we can never see through the cat's eyes, research data do give us some insight into how cats see in daylight and darkness. Experiments indicate that light stimulates some retinal receptors to "turn on," whereas it causes others to "turn off." In cats, these two types of light-sensitive structures create a central core of "on" receptors surrounded by a ring of "off" ones. This arrangement allows cats to see the edges of objects clearly in bright light in spite of their relative lack of detail receptors. To get some idea how such a system works, observe the edge of this book: Now curl your fingers to make a tube and observe the same edge again. Notice how both the blocking and the relative darkness produced by your hand makes the edge of the page more distinct. Although this doesn't give the amount of detail necessary to specifically identify or read the book, you could certainly locate it given this kind of visual information. Similarly, because the cat uses its vision to locate more than specifically identify objects, this system functions quite well.

Despite the speculative nature of such considerations, human appreciation of such differences can benefit our relationships with our cats.

The Body Language and Emotion of Cats

The instant we assume that our cats see things the same way we do, we begin to interpret their body-language displays erroneously. By projecting a human-eye view on the cat, thus denying it (and ourselves) its own unique orientation, we rob ourselves of a potentially much richer experience. If Lisa McDermott calls her cat stupid because "he can't see his ball right there under the couch" or because "he takes off across the yard after nothing," her beliefs may cause her cat no physical harm. On the other hand, how does believing her pet stupid versus perfectly normal affect their relationship?

The Ultimate Night-light

For the nocturnal predator, feline ocular anatomy fills the bill perfectly. Imagine sitting in your darkened living room. How much color can you see? How much detail? Now imagine that your life (or next meal) depends on your ability to detect any changes that occur in the dull brown chair far across the room. Which change would be the easiest to spot:

- A change in color from brown to gray?
- Double instead of single stitching around the pillows?
- A sudden movement of the chair two inches to the right?

As light dims, color and detail become less important, and movement becomes correspondingly more critical. To the hunting cat, the mouse's color or sex makes no difference. Movement does. When the food stops moving, smell and taste come into play, further negating a need for color and detail vision.

In addition to a light-reflecting tapetum and a preponderance of motion-sensitive rods, the highly flexible feline pupil contributes its share to the cat's superior nighttime vision. The pupil can dilate to produce an opening almost twelve millimeters (almost one-half inch) wide, compared to the maximum human width of approximately eight millimeters. Consequently not only does the cat enjoy a mechanism for utilizing available light, it collects light more effectively too. Of course it's impossible to talk about dilated feline pupils without mentioning the opposite extreme, that barely perceptible slit that blocks out light on a bright sunny day. Why isn't the cat's constricted pupil round like most other mammals, including humans? The answer lies in an anatomical paradox shared by all mammals. Stimulated rods and cones send their messages to the optic nerve, which connects the eye to the brain, which, in turn, interprets the visual data. However, while the area where the optic nerve attaches to the back of the eyeball (called the optic disk) certainly does a marvelous job at *transmitting* data, it contains no receptors itself. Therefore any light striking the optic disk creates no visual image because no rods and cones occupy that area— although this effect is lessened somewhat because the receptors are more densely packed around the optic disk than in other areas of the retina.

Unlike the optic disk of most other mammals, which sit ventrally, the cat's occupies a central position. If the feline pupil constricted to form a centrally located round opening like ours, only the nonreceptive optic disk would be exposed to light, rendering the cat functionally blind in bright light. However, thanks to the unique pupil that constricts vertically cats, too, can see in bright light in spite of this anatomical variation. Furthermore, and again unlike most other mammals, the cat's reflective tapetum completely surrounds its optic disk, guaranteeing the even diffusion of light necessary for optimum stimulation even when the constricted pupil limits the incoming light to only a fraction of the total retinal area.

While the waxing and waning feline pupil provides us with an awe-inspiring array of anatomical and physiological diversity in response to light, it also provides valuable clues to the cat's mood. If the eyes are the "windows of the soul," then narrowed pupils signal a cat on the offense, while fully dilated ones indicate a cat in a defensive mood. Despite a dearth of research in this area, it seems logical that a cat about to attack would profit from a narrower field of vision, which would enable it to block out any visual distractions. Moreover, because we know the visual receptors increase in number closer to the optic disk, narrowed eyes may enhance perception of detail or fine motion,

THE BODY LANGUAGE AND EMOTION OF CATS

information crucial to targeting an attack successfully.

On the other hand, the defensive cat, who wants more than anything else to be able to respond to changes in a threatening situation, needs to collect as much data as possible. Because it can't afford to limit its concentration to one area, it dilates its pupil in an effort to detect movement over a wide area. Imagine yourself a 115-pound woman evaluating a four-year-old terrorizing his playmates with a stick versus a drunk six-foot, 200-pound knife-wielding male. In the former instance, you concentrate all your effort on the child, seeking the best opportunity to grab and disarm him. In the latter, all your effort goes into finding the safest, fastest route of escape.

Every time I think about these behaviorally linked pupillary displays, I'm reminded of the association between vertical and horizontally slit pupils in poisonous and nonpoisonous snakes: I can never remember which kind of pupil goes with which kind of snake. Luckily, however, offensive and defensive cats flash sufficient other body-language signals. Once we learn how to read all the signs, the chances of being surprised by a sneak feline attack drop dramatically. For now, simply remember that a cat displaying very dilated or constricted pupils under normal light conditions is telling you that something isn't right. If the cat appears totally relaxed, it could be suffering from a medical problem, and a trip to the vet is in order. If the cat appears the least bit agitated, back off and give it time to settle down.

FROM THE KITTEN'S EYE VIEW TO THE CAT'S WORLDVIEW

Some fascinating experimental results have provided further hints about the exquisitely balanced feline visual system. In one experiment researchers placed young kittens in boxes, the interiors of which were painted with vertical lines; they put another group in boxes decorated with only horizontal lines. If the kittens remained in these visually restricted environments during a short, but critical, imprinting period of their kittenhood, the results were astonishing. Those exposed to the vertically lined environment could deftly negotiate mazes of chair and table legs, but never jumped up or down to other levels. Those placed in the horizontally lined containers readily jumped from level to level but stumbled into vertical chair and table legs as if blind to them.

When we consider the way age and experience literally program the cat's vision for life, and couple this with the data regarding the far-reaching effects early feeding experiences wield over food preference, we can understand why behaviorists and veterinarians urge people to make every effort to find *feline* foster mothers for orphaned kittens. This also explains why these experts worry about kittens raised in restricted environments, such as cages—be they in the most sophisticated cattery or in a dank and dingy pound. Only kittens who enjoy the tutelege of their own kind in a sufficiently varied environment fully develop the perceptual responses we associate with normal, healthy feline behavior.

THE OCULAR COVER-UP

Compared with humans and dogs, cats basically lack eyelashes, which seems odd considering their nocturnal and stalking habits. Because eyelashes help protect sensitive ocular tissue from dust and other relatively large particles, surely a cat's lack of such protection indicates Mother Nature slipped up somewhere along the evolutionary line. Actually, cats enjoy a much more effective protective mechanism: the third eyelid or nictitating membrane, which elevates any time the eyeball is pushed or pulled back into its bony socket. If you ever noticed this white membrane across your sleeping pet's eyes, you may have thought that the cat's eyes had "rolled back" into its head. However, you actually saw the third eyelid, not the white (or sclera) of the eye itself. When the animal awakes, the third eyelid sinks into position in the lower corner of the eye next to the nose; but once again, owners often think the eyes have "rolled down" again rather than that the third eyelid has slid away to reveal the eye in its normal position.

Unlike that of other mammals, the feline nictitating membrane responds to voluntary control in addition to passively protecting the eyes.

THE BODY LANGUAGE AND EMOTION OF CATS

Tiny muscles enable the cat to raise the third eyelid independent of any action of the eyeball. Why would such an adaptation benefit the cat? Consider these facts:

- Relatively speaking, cats are small predators who stalk their prey close to the ground.
- Cats are nocturnal.
- Cats possess minimal detail vision.
- Mice and other rodents tend to live in thickets or other structurally complex areas rather than smooth, open spaces.

Most commonly, the third eyelid functions as a shield, which the cat elevates and retracts at will to protect its eyes during the hunt. To test this theory, I crawled through high grass on my stomach, stalking Maggie, who was herself stalking a garden snake. I tried to stay beside but a bit ahead of her so that I could see her eyes, a position that offered the additional exciting prospect that the snake might veer and slither across my own path at any instant. As I watched the cat at work, I was struck by the complimentary relationship between the third eyelid and the slit pupil. Given the cat's elongated rather than circular pupil, it can see "above" the elevated third eyelid. In such a way it relinquishes little of its visual field to gain the extra ocular protection.

One or both third eyelids also elevate in response to ocular disturbances, such as foreign bodies, infections, ulcers, tumors, nerve damage, and even some medications. Likewise cats with chronic or wasting diseases will also often expose their third eyelids. Although perhaps most commonly associated with severe illnesses, such as feline leukemia or infectious peritonitis, a significant number of cats also exhibit this behavior when experiencing intermittent or chronic bouts of diarrhea associated with food allergies or parasitism.

Because two different mechanisms can produce the display, we can't always be sure about any exact causes if no other signs are present. On the other hand, a disease that results in weight loss and dehydration could certainly result in "sunken" eyes that trigger the passive elevation of the third eyelids. Similarly, anything damaging the ocular muscles and nerves could also interfere with normal membrane function. Finally, because the voluntary mechanism does exist, we can't rule out the possibility that a cat who feels ill or whose eye hurts may choose to block out certain visual stimuli, much as a person with a headache or eye infection will often shield his or her eyes from light. Regardless of which mechanism causes the display, the consistent presence of the third eyelid provides owners with a valuable warning sign they should not ignore.

HEAR YE, HEAR ME

Few dispute the critical role hearing plays in the survival of the nocturnal predator, but testing such an animal's hearing under *natural* circumstances has proved an elusive task. Even if we could do it easily, such findings might not necessarily hold true for your Havana brown, who, like his father, grandfather, and great-grandfather before him, sleeps the night away and limits his foot-related auditory responses to the sound of the can opener and his name. However, one thing for sure, cats do hear a lot more than people do, responding to sounds in the fifty to sixty-kilohertz (thousand cycles per second) range, compared with a human range that peaks at eighteen to twenty kilohertz. When you consider that we humans label anything above twenty kilohertz "ultrasonic," our cats live in a world of sounds far beyond our comprehension.

The cat's collection and analysis of this far-greater amount of auditory data begins with the pinnae, or external ears, which it can direct much like a dish antenna to funnel maximum sound waves into the canal leading to the middle and inner ear. Although dogs with erect ears also use their pinnae in this fashion, like so many feline movements, the cat's manipulation appears much more delicate and precise. Any sound-collecting advantage possessed by flop-eared dogs obviously diminishes to some extent because of the mechanical obstruction created by the pinnae themselves; and of course we humans with our fixed pinnae must rotate our whole heads if we wish to pinpoint a sound source.

What survival advantage do the highly moveable pinnae confer upon the cat? Given the feline preference for hunting in limited light, its minimal detail vision, and its predilection for prey that itself communicates in the ultrasonic range and has motion-sensitive vision, a stalker must make a minimum of noise and motion in order to locate its prey successfully and avoid detection. Similarly, an ability to locate the source of external cues with minimal body-language expression offers valuable protection for a predator, which may, in turn, be preyed upon by others.

Observe your cat attempting to locate and evaluate an unfamiliar sound. Unlike humans, who turn their heads from side to side and may even move their whole bodies in the direction of the sound, cats usually remain still, save for precise flicks of the pinnae.

READING EAR LANGUAGE

Because cats rely so much on their hearing, the position of the feline pinnae provide us with valuable clues as to how the cat takes in its surroundings. For example, an alert, confident cat who considers you nonthreatening holds its ears erect and facing toward you because it's willing to collect more sound data from you and the area immediately around you.

If the cat decides that you represent a threat it must deal with offensively (that is, get you before you get it), the pinnae will swivel so that the insides of the ears face to the side. Having decided to attack and located its target, it now uses its hearing both to help select the proper moment for attack and to protect itself from surprise attack from the side and rear.

The cat who considers you a threat but wants to be left alone will fight only to defend itself; it holds its ears tightly against its head. Why a cat who considers itself endangered would block sensory input to its most finely tuned sense seems a most foolish behavioral contra-

Feline Sensations: Reading the Mind of the Demon-God

diction. We know that animals in such a situation exhibit fully dilated pupils and therefore become more motion-sensitive, but why in the world wouldn't they want to hear?

Several answers come to mind, but fear seems the most logical explanation: The defensive body-language display with the body pressed tightly to the ground, ears flattened, and pupils dilated best enables the animal to assume the freeze response, which in the wild communicates, "I don't want to fight. I can't or don't want to run." To maintain this position the cat may need to block out as many sounds as possible lest it panic. This explanation reminds me of my own love/hate relationship with fireworks when I was a child. I'd sit huddled between my parents watching the skyrockets flaring and bursting, eyes big as saucers and hands firmly clamped over my ears; I wanted to see the display, but I didn't like the scary noise at all!

The second possible explanation centers around the physical advantages offered by the flattened ears. If the cat refuses to fight but fears attack, it may flatten its ears not only to help make itself look smaller and less threatening, but also to deny its assailant these two convenient teeth-holds. Because cats are solitary creatures and prefer to avoid confrontation, little data exist on the nature of true feline hostility. However, veterinarians routinely see cats with head wounds, and the presence of the protective shields in intact toms certainly gives credence to the idea that this area is a preferred target.

A third possible explanation relates to the second. If the defensive cat expects some sort of attack, but not a life-threatening one, it may cover its ears to protect its most critical sensory apparatus. Remember that intraspecies confrontations most frequently terminate when one individual behaviorally rather than physically manifests its dominance or superiority. Therefore, defensive animals who don't expect to be killed may assume positions designed to limit any damage to nonvital parts.

Again, although we can't know exactly what transpires in the feline mind when these various pinnae alignments occur, we can use these

THE BODY LANGUAGE AND EMOTION OF CATS

signals to avoid provoking or further frightening a threatened animal and perhaps save ourselves from a few scratches and bites to boot.

HEARING THE UNHEARD

Scientific evidence is gradually accumulating to support what cat owners have known all along: Cats communicate in ways beyond human comprehension. I'll never forget the shock of recognition that swept over me when I read Paul Gallico's description of the silent meow in *The Abandoned*. Physiologically it makes sense: Why would a queen bother communicating with her offspring in the paltry range of human hearing when her young can hear so much more? But alas, that perfectly reasonable line of thinking led me to a more startling question: Given that its prey as well as its own species communicates in the ultrasonic range, why do cats bother learning to communicate in the human range at all?

On several occasions I've met cats or kittens who were raised away from people until later in life. If these cats wanted food, water, or to go out, they sought out their owners just like any other cat and meowed —but silently. The cat looked like it was meowing, acted like it was meowing, but not a sound could be heard.

One such cat and her kitten temporarily moved in with me. The reason I support the view of the silent meow even though I could uncover little scientific explanation for it springs from my observations of this cat and her kitten. On several occasions I observed the queen awake from her nap, stretch, and begin grooming herself. Once she completed her toilet, she'd emit the silent meow. As if by magic, I would hear a thump, then the scurrying of little feet as the kitten whisked into the room to touch base with Ma.

On another occasion I found myself in one of those wretched "Do as I say and not as I do" situations: I wound up bottle-raising a two-day-old kitten, Ray, born to a stray who lived in an abandoned warehouse. During the first few weeks, the kitten experienced little interaction with other cats. Then he lived with a female, who gave birth to a litter when Ray was about six months old. Ray the Stray never uttered a sound until those kittens began making audible meows; then he took a stab at imitating them. It sounded as if he were attempting to play a musical instrument for which he lacked any natural ability. He never got the hang of it and settled instead for a semicroak, which he emitted only when all else—including throwing his whole body against or on top of me—failed. Now we only have one cat, the aforementioned Maggie, who is from that litter, and she, like her mother and Ray, uses audible communication only when all other forms fail to yield the desired human response. The one exception to this rule will strike a responsive chord with many owners: If I don't open the door quickly enough, particularly if she's outdoors on a cold and rainy day, she enters meowing loudly in what can only be described as feline complaining—or worse.

While the ailurophile in me might speculate that cats consciously choose to communicate in a sound range accessible to human ears, we can't overlook the role those raucous, vulgar yowlings play in normal feline mating behavior. A female belonging to a solitary species needs some way of attracting suitors. True, the female secretes potent sex hormones whose scent attracts males; but the back-fence yowling may act as a backup alerting system, as well as enabling the animals to locate each other more precisely. Why don't cats give us humans a break and do their caterwauling in the ultrasonic range? They would still hear it—and maybe even more clearly—and you and I could get our sleep.

Again, several possible answers leap to mind. First, the range itself may identify the meaning of the message. For example, if queens use the ultrasonic range to communicate with kittens and all cats use it to detect rodents, mating calls in that same range might either be lost in the confusion or interfere with these other critical functions. If mating calls occur in a completely different range, those who aren't interested in mating—females who've already been bred, immature felines of both sexes—could simply ignore these sounds. Compare this to listen-

THE BODY LANGUAGE AND EMOTION OF CATS

ing to the radio and completely ignoring the sound of the automatic washer because you want to hear the latest weather report so that you know whether or not to wear a sweater. On the other hand, if you suspect that the washer may be overloaded, you focus on the range of its sound and ignore the radio.

Or perhaps sounds in the lower ranges escape the notice of predators. Although humans can hear these pitiful screams painfully well but not a bit of what goes on in the ultrasonic range, maybe some animals experience the reverse, finding ultrasonic sounds much easier to locate and identify than those within human earshot. Although the bred female does experience a much shorter heat cycle than the unbred one, we're still talking about four to six days of what surely amounts to "Come and get me" invitations to predators as well as mates. The fact that females aren't routinely slaughtered during heat even though predators may be in the area would seem to indicate that the latter might simply be overlooking the commotion. Although this might appear a bit farfetched, recall how many times your cat may appear totally oblivious to the sound of its name but perk right up when you whistle to make a particular chirping noise. If the cat was intent on sounds in another range, it simply didn't hear you until you communicated to it in that range.

Finally, perhaps cats make all this noise in the range of peak human sensitivity because of, and in response to, people. While this may seem facetious, we actually know little about how cats live in the wild. To collect accurate data, we must observe the cats ourselves or set up instruments to do the job for us; either way, we inevitably leave behind all sorts of sensory clues. Therefore regardless of what data we collect, we can't be sure that they reflect what happens when we and/or our equipment aren't present.

Although dog people might have trouble with such an obscure concept, Nobel physicist and philosopher Erwin Schrödinger once posed a fascinating dilemma that has come to be called "Schrödinger's Cat." The problem proposes the following: If we place a cat in a box and

Feline Sensations: Reading the Mind of the Demon-God

cover the box, is there any way we can be absolutely sure what the cat does in the box without uncovering it? Obviously not. But as soon as we uncover the box to peek inside, we disturb the cat. The same holds true for any means of observation we may use; these can only tell us how the animal behaves in the presence of the observer and any instruments, which may or may not be the same way it behaves when alone.

While "Schrödinger's Cat" may strike some as a useless mind game, anyone who wants to understand cats intimately must think about it. Consider this tale of scientific data collection. Back in the 1940s, two scientists, E. R. Guthrie and G. P. Horton placed cats in special boxes designed to test their problem-solving ability. The experiment involved each cat learning to lift a rod that would then open the cage door. Repeated observations showed that the cats always displayed a particular sequence of body-language signals before they lifted the rod and opened the door: They rubbed their faces and bodies against the door, circled, and finally lifted the rod. This experiment became a classic, especially since it adhered to the most objective and stringent scientific criteria of the time. The conclusion? Cats somehow *need* the complex body-language display to enable them to open the door. Ailurophiles cheered because such scientific proof of the cat's almost mystical ceremonial behavior gave credence to their own worshipful beliefs that cats displayed behaviors far beyond the capacities of other animals or even human understanding.

Then in 1979, B. R. Moore and S. Stuttard repeated this famous experiment. They, too, observed the complex behavioral sequence; but another, even more fascinating fact emerged. The cats only displayed the behavior *when people were watching*. If the people went away, the cats slept contentedly; but as soon as someone entered the room, they resumed their ritual dance—regardless of whether they were in the cages or not.

Bizarre? Think about it. Head and body rubbing, circling, touch. Moore and Stuttard recognized that complex behavior for what every cat owner knows it to be: the typical display cats use to greet people!

THE BODY LANGUAGE AND EMOTION OF CATS

So what does all this have to do with sound? Quite simply, the fact that the sounds cats utter in the human range could in fact be responses to human presence. Consider the charming "talk" of the highly vocal Siamese, whose coloration has made it a natural for human protection and companionship. Some animal behaviorists postulate that people selectively bred these cats for that very sound because of its human-baby-like quality. In other words, the Siamese evolving to survive in a human world needed to develop skills or forms of communication within the range guaranteeing maximum human perception and response.

The question then becomes: Did these nonultrasonic skills result from changes in feline anatomy and physiology responding to natural selection and the cat's association with humans, or did we force these changes on the species via selective breeding because we wanted to align them better with our own perceptions? We'll probably never know the answer, although personal experience leads me to believe it's a combination of the two; I find it hard to accept that cats were ever completely passive participants in their own evolution.

Let's consider another example. I have had females of breeding age that were strictly house cats, as well as those that moved freely in and out in large, fairly secluded areas. I've also worked with clients whose cats belonged to one of these two categories. Without a doubt, owners of strictly house cats or those living in densely populated areas complain about their pets' vocalization much more than do their rural or reclusive counterparts. In fact, those whose cats can romp freely over large areas rarely if ever even realize their females were in heat until they become obviously pregnant.

True, we can say that house cats continue yowling because their pleas have gone unheeded. But even after these cats have been bred, the yowling often continues, albeit at a lower level. And what about all the feline songfests in Maple Acres, where every house in the subdivision contains at least one cat? There's no need for these cats to call each other when they can see, smell, and even spit at each other. But

Feline Sensations: Reading the Mind of the Demon-God

perhaps these cats really don't make more noise at all; maybe those who live in these areas just hear them more readily because they're so close by.

Another example of vocalization in response to human association involves the peculiar form of call made by cats who've lost their sight. On several occasions I've examined cats whose owners' only complaint concerned the cat's sudden tendency to yowl as it moved through the house. In most instances, these were older animals who otherwise behaved normally. The offending cries differed noticeably from other forms of vocalization the owners previously recognized, but because the cats seemed normal in every other way, the owners usually didn't seek medical assistance until the behavior began to jeopardize the relationship. In all cases but one, the only change I could detect was a loss of vision, usually due to age-related changes in the eye. We know that bats and sea mammals utilize echolocation, an orienting system whereby they bounce sound off objects to locate them. Is it possible that vision-impaired cats use a similar system? While the particular sounds produced signaled something "wrong" to the owner, none of them noticed their cats bumping into things. However, once aware of the cat's visual impairment, they quickly realized their pets no longer engaged in activities that required vision, but rather preferred activities and toys involving sounds or scents (catnip mice with bells). Moreover, these cats often developed a strong attachment to particular resting or sleeping places, especially bedding consisting of some article of the owner's clothing. In such a way they apparently compensated for the loss in vision by redefining their visual worlds in terms of sounds and scents.

One cat who didn't fit this pattern was a young female who suddenly lost her hearing. In this case, the cat suffered a severe blow to the head when a garage shelf collapsed on her. The cat fled and returned shouting the next day. The only apparent abnormality? Deafness. She stuck to her owners like glue and produced loud, piercing cries for several days. Soon these occurred only when the owners left the cat's field of vision, and the cat went on to live a long and happy life.

It certainly seems possible that the cat initially vocalized in response to the distress and disorientation that accompanied its sudden loss of a critical sense. Because the owners sought out the cat and comforted her when she made this display, the cat linked their behavior to this peculiar cry; and even though she adapted to her deafness, she continued to use that cry to bring her owners to her when she needed help. This in turn convinced the owners that not only had their pet compensated for her hearing loss, but she had also learned how to alert them if she needed them. In this particular case, the cat never abused the relationship by letting out those "Come here!" attention-getting calls common to many cats, especially the more vocal breeds. It was as if this cat fully understood the value of this special form of communication and would never jeopardize it by crying wolf.

Such unsettled issues seem to say that cats may maintain two separate standards and life-styles—one reflecting their relationship with their natural environment and one reflecting their relationship with people. Regardless of whether the cat functions better in the ultrasonic or human range, it can and does function effectively in both. And although genetic manipulation may have predisposed certain breeds to be more vocal, only the most naive would designate animals possessing this characteristic as either superior or inferior to the silent stalkers of the night. We can never know exactly how much more or less a cat hears than we do, or what those sounds mean to it; but we can and should cultivate the patience, knowledge, and acceptance that will allow our cats to function in both worlds of sound harmoniously.

In our brief discussion of feline senses, we seem to have raised as many questions as we answered about how and why our cats behave the way they do. However, rather than feeling frustrated by our inability to resolve the paradoxes, clarify the inconsistencies, or wave away the ambiguities, we can embrace them in all their wonderful and contradictory beauty. Lacking rigid definitions, we can allow unique

Feline Sensations: Reading the Mind of the Demon-God

interpretations of our own cat's behaviors and develop a unique relationship to complement those interpretations. In the next chapter,
we're going to delve into the human response. Are you an ailurophile
all of the time, or do you guiltily acknowledge a few times in your
life when you would have gladly given Felix a one-way ticket to the
moon?

4

THE HUMAN RESPONSE: RUNNING THE EMOTIONAL GAMUT

*O*NLY faint light tinges the morning sky when the Palmieris' Himalayan, Lotus, gracefully leaps into the brand-new crib to check out the latest addition to the household, one-week-old Anthony Palmieri, Jr. When the proud papa finds Lotus in the crib a short time later, he sees red: "Get the hell out of there, you sneaky beast!" The cat scurries through the door just inches ahead of the flying shoe. Naturally Tony doesn't believe that old wives' tale about cats sucking the breath out of babies. On the other hand, it infuriates him that his expensive purebred cat so flagrantly ignores his wishes by doing something so common as to sleep in a forbidden area. He expected that kind of behavior from the ragtag group of street cats he grew up with, but most certainly not from an animal with Lotus's credentials.

Next door the Palmieris' neighbor, Sharon Fairbanks, looks deep into the eyes of the orange tabby enveloped in a cocoon of bedclothes perched on her chest. "Pretty Percy, kissy, kissy, kissy," she coos lovingly. "Mummy can't imagine what she'd do without her Percy-love to keep her company." Sharon suddenly aborts the rubbing and petting that has Percy rolling in obvious delight among the blankets. "Damn, I forgot, I have to get your fish tonight!" Not only does this recollection snuff out her plans to come home immediately after work and treat herself to a leisurely bubble bath and quiet dinner with a

The Human Response: Running the Emotional Gamut

friend, it also means she'll be thrust into the very predicament she hoped to avoid—negotiating rush-hour traffic in what forecasters promised would be freezing rain changing to snow.

If such seemingly unemotional, straightforward topics as feline anatomy and physiology can ignite a wide range of human emotions, surely something as emotionally charged as body-language expressions can lead us into a hopelessly tangled web of feelings. If so, why even bother delving into the web and all its sticky questions? After all, for every million cat fanciers, there must be at least a million different responses to any particular feline display.

Actually, in spite of all our diverse opinions and idiosyncrasies, we humans essentially adhere to the laws of nature, the primary one of which demands orderliness. Like all other living beings on this earth, our actions and reactions adher to a certain logic, predictability, and orderliness, even when it comes to our responses to feline mysteries. However, we often get so caught up in a specific interaction that we either don't have or don't take the time to evaluate that particular event with any kind of clinical objectivity or with any view toward how it affects our total relationship with our pets. Unable to stand back and see the forest in all its splendor, we spend so much time bumping into individual maples, oaks, and birches that we quickly find ourselves lost deep in the woods. Suddenly Tony Palmieri realizes that his singular interactions with Lotus have deteriorated into armed combat; he subconsciously begins rating Lotus's behavior (good or bad), then scoring it (*how* good, *how* bad). As the relationship between owner and cat gets lost in this forest of feelings, Tony actually finds himself hoping for "the last straw," that climactic event that might relieve the unbearable tension between him and his cat.

This certainly isn't what Tony had in mind when he celebrated his wife's pregnancy by surprising her with the Himalayan kitten six months ago. On that golden day Lotus symbolized the Palmieris' new family status. However, now she seems to be more intent on tearing the fledgling family apart. Many marital disagreements center on Tony and Kim Palmieri's views of Lotus's behavior, a trend that could easily accelerate now that another Palmieri has come into the picture.

When we compare the opposing views of ancient Egyptians and medieval Europeans, we're tempted to chalk them up to unsophisticated foolishness, something we would never do. Sharon Fairbanks directs the nursing staff at one of the nation's top teaching hospitals; the idea of embarking on a lusty pilgrimage down the Nile to honor Bastet would make her snort with disgust. On the other hand, she does drive miles out of her way to purchase fresh fish for Percy every week. Tony Palmieri's skill and patience dealing with troubled teenagers has won him numerous civic awards; even his worst critic would never stoop to call him a barbarian. But what *do* we call a person who fantasizes about diabolical ways to get even with a perfectly normal cat?

BALANCE AND PERSPECTIVE

A favorite wall hanging that inspired me during my final years of veterinary college and early years of practice contained two words handwritten in elaborate caligraphy: *Balance* and *Perspective*. As we explore the many forms of human response to feline body language, and as we interact with our own individual pets, we need to call upon these two concepts as standards for evaluating ourselves. Although we'll be discussing specific types of owners whose behaviors may appear exaggerated or stereotyped, bear in mind that most owners exhibit some form of such behaviors at one time or another. We vacillate as we try to establish an arbitrary point at which we feel our human needs balance those of our cats. If we move too far to the feline end of this scale, we feel manipulated; if we move too far to the human end, we feel selfish and guilty. The key to a happy medium? The proper perspective, which, like distance from the individual maples and oaks, permits us to view the whole panorama of experience available.

Although all pet owners find their relationships with their pets precariously balanced from time to time, few consciously set out to create such an instability. Sharon didn't intend to make the fifty-mile fresh-fish run every week; Tony didn't buy Lotus with the idea that he and his cat would participate in a daily battle of wills. Because we rarely define what we mean by *balance* specifically enough, we begin to accept one side of the scale or the other as a normal—and unchangeable—part of cat ownership. Ironically, while no owners would define either themselves or their cats as physical or behavioral statues, we do tend to respond to situations quite rigidly. Although Sharon may not enjoy the weekly trek to Fairfield's Fresh Fish Emporium, Percy's obvious delight with his fishy fare convinces her that she must. In such a way, these trips become instrumental in maintaining the relationship between Sharon and Percy. If Sharon eliminates the fish from Percy's diet simply because she's tired of the long drive, chances are she will feel that she has deprived Percy of a quality relationship.

Of course, you and I could objectively argue that Sharon's love for her cat and his love for her could surely survive canned-tuna cat dinner. Yet, despite our ironclad logic and Sharon's ability to evaluate, change, and adapt her complex interactions with her hospital staff and students, she can't muster the confidence to act that way with Percy.

Regardless of whether you consider your relationship with your pet ideal or not, bear in mind that it represents a balance that you've struck between your needs and your cat's. Keep reminding yourself if necessary that it only represents the balance of your needs and your cat's *at this particular time.* And try to recognize that both your own and your cat's needs can and do change; the ideal or "right" human or feline response to a particular event on one day may not be appropriate on another. Right or wrong depends on how the response affects the relationship in the long run.

The second quality, *perspective*, not only provides objective distance from each interaction, it also helps us fit a given interaction into our overall relationship. To be sure, one rotten apple can spoil the whole barrel, but only if we leave that apple in the barrel. If we remove it the instant we spot it, chances are we can limit the damage.

THE BODY LANGUAGE AND EMOTION OF CATS

So, while we all aim for a lifelong mutually rewarding and beneficial relationship with our cats, each incident carries its own potential to enhance or undermine that greater goal. The perspective we bring to each event, whether we allow it to elate or depress us, can wield tremendous power when it comes to forging a solid human/feline bond. If Tony Palmieri chooses to disregard the positive aspects of Lotus's displays and concentrates instead on what he considers her major deficiencies, this negative perspective will surely undermine a stable balance between him and his pet. By choosing to recognize only what Lotus does wrong, he surrounds his entire relationship with Lotus with a negative aura.

As we discuss the various types of human responses, evaluate your own in terms of balance and perspective. If you catch yourself taking offense at or feeling defensive about a particular orientation, examine these feelings and evaluate them honestly and objectively. You may feel very strongly about the "rightness" or "wrongness" of Sharon's or Tony's relationship with their pets, but bear in mind that there's a bit of Sharon and Tony, a bit of ailurophile and ailurophobe, in each of us. Those who would deny this paradoxical duality deny themselves the opportunity to participate in the full richness of human/feline relationships. And those who deny themselves objectivity deny themselves the opportunity to change a relationship from less-than-perfect to ideal.

THE VIEW FROM WITHIN THE TEMPLE

Before we launch our discussion of ailurophilia, let's make sure we're all talking about the same thing. That the words *ailurophilia* and *ailurophile* crop up in conversations among cat owners, yet don't appear in my standard dictionary, attests to the exotic nature of the relationship between people and their cats. It also raises the possibility that, lacking any formal accepted definitions, people may assign whatever meanings strike their fancy. My quest for a clear definition of these elusive terms led me to the pages of that professional classic, *Dorland's Illustrated*

The Human Response: Running the Emotional Gamut

Medical Dictionary. Although my study of the relationship between human and feline should have rendered me immune to surprise regarding any paradoxes I might encounter, I must admit that my view of lexicographers as a dispassionate crew caused me to let down my guard. The venerable *Dorland's* informed me that ailurophilia comes from the Greek *ailouras* ("cat") and *philein* ("to love"). So far, so good; obviously, as any cat lover knows, an ailurophile is someone who loves cats. But unless your physician happens to be a cat lover too, don't brag about being an ailurophile; according to *Dorland's*—and in spite of the word's origins—an ailurophile is someone having a "*morbid* or *inordinate* fondness for cats.*" In other words, ailurophilia is an unnatural, unhealthy, "sicko" orientation.

So, many standard dictionaries say we don't exist, and medical dictionaries say we're nuts. What's a cat lover to do? First, we can define our own terms. The word *ailurophile* probably entered common catfancy parlance thanks to a literal translation made by some cat lover unmindful of the pejorative medical definition. Of course, that literal translation is, in fact, linguistically purer and more accurate than the medical one. However, if you proclaim your passionate ailurophilia at a cocktail party and sense certain individuals edging away from you (or toward you carrying black bags and wearing worried expressions), don't be surprised.

For our purpose, we're going to use the term *ailurophile* to describe those people who, like the ancient Egyptians, believe that their cats can do no wrong. Or, perhaps more correctly, they see their cats as so different that any feline misbehavior must be accepted because the distance between them precludes any changes. Cats will be cats. When Percy gobbles down his fresh fish, Sharon concludes that he needs it— "Cats always know what's best for them"—and not providing it would be negligent. It never crosses her mind that Percy might pursue a behavior not in his best interests, or that her own feelings merit equal consideration when fulfilling his needs.

This tendency to revere cats by ascribing to a form of love that surpasses all human understanding works well as long as it works well. As long as whatever Percy does (or doesn't do) doesn't exceed the limits Sharon sets for a balanced relationship, the two of them can coexist satisfactorily. However, if Percy does something Sharon considers incomprehensible and wrong, her ailurophilic orientation leaves her no option but endurance. For example, having decided to honor Percy's predilection for fresh fish from Fairchild's, Sharon feels compelled to make the weekly trip. When the weather's fine, when a favorite companion accompanies her, and when the service at Fairchild's is flawless, Sharon truly enjoys ministering to Percy in this manner. However, when freezing rain turns the trip into a fifty-mile solitary nightmare interrupted only by an unpleasant exchange with a surly clerk at the fish market, Sharon's attitude toward the ritual alters dramatically. Rather than feeling benevolent toward Percy, she loathes him for putting her in this predicament. But because her ailurophilic definitions allow Percy to do no wrong, Sharon winds up feeling guilty about her negative feelings: How can she possibly hate Percy for being a "normal" cat?

Although ailurophilic owners come from all walks of life and tolerate all sorts of feline behaviors, they do share one common characteristic: lack of confidence. In confident, loving relationships one participant need not obtain happiness at the expense of the other. If cat owners believe that they must suppress or feel guilty about any negative feelings rather than confront them openly and honestly, perhaps they do deserve *Dorland's* definition. Anyone who thinks that he must make sacrifices to maintain a cat's friendship deludes himself and risks damaging the bond so necessary to a healthy, loving relationship. Although some owners say that they'd do anything for their cats, thereby implying their willingness to take total responsibility for any course the relationship takes, in reality they see themselves as slaves at best, victims at worst. In such a way, their intense devotion can actually turn into its opposite: A lack of confidence leads to feelings of manipulation; manipulation leads to anger; and anger leads to guilt.

The Human Response: Running the Emotional Gamut

LOVE ME, LOVE MY CAT

Moving in from the ailurophilic extreme of the human-response scale, we encounter the more common anthropomorphic orientation. Unlike the ailurophilics, who imbue their cats with qualities beyond human comprehension, those ascribing to an anthropomorphic view interpret feline behaviors in purely human terms. Whereas Sharon Fairbank's ailurophilic fifty-mile fish pilgrimage reflects her belief that "Percy knows what's good for him," an owner ascribing to the anthropomorphic view would say, "I know just how Percy feels. That frozen fish from the supermarket tastes just like cardboard."

Anthropomorphic responses generally arise when we lack the experience, knowledge, or confidence to handle a situation in a more species-specific manner. Lack of experience and knowledge commonly afflict first-time owners simply because their understanding of normal feline behavior may be incomplete or laced with mythology. When confronted with problems, these owners apply a familiar standard: human behavior. To see how this works, let's examine the roots of Sharon's anthropomorphism.

Sharon entered adulthood with two goals: to be self-supporting and to own a cat, despite—or perhaps because of—the fact that her mother had never permitted her to have a pet when she was a child. The combination of her potent lifetime fantasies about cats and her complete lack of experience prime her to respond to Percy as she would to a child whenever she sees the slightest similarity between his behavior and that of a human. While she may relegate his penchant for chasing mice and spraying urine to the realm of the gods, she can readily relate most of his eating and sleeping habits to her own.

Tony Palmieri also suffers from the new owner's lack of experience, but in a more insidious way. Although Tony has had cats rubbing against his ankles throughout his life, Lotus is his first purebred, the first cat he actually bought. In his mind Lotus's $150 pedigree somehow makes her different and more special than the common shorthairs and strays that populated his life in the past. As a result he spends

THE BODY LANGUAGE AND EMOTION OF CATS

more time observing and analyzing her than he ever dreamed possible, always expecting her to be smarter, better behaved, more affectionate, more aloof, more active, and more mellow than her barn-bred counterparts. Because of his cat's pedigree and his financial investment, when Lotus does something Tony finds incomprehensible or negative, he responds to her as if she were a person. Show me an unknowledgeable cat owner who paid a lot of money for a pedigreed cat and I'll show you a person who treats his or her cat like a spoiled, but beloved, child. In fact, one wag at a cat show once joked that the only difference between a cat and a baby is about $250—the cat costs more!

Most people accept eccentricities in cats much more readily than they do with any other species. Even those who disdain anthropomorphism as appalling or "sick" will often blame the cat for the aberrant relationship, whereas they wouldn't hesitate to blame the owner in a similar human/canine interaction.

Such widespread acceptance and tolerance of anthropomorphism does spring in part from the cat's very nature. Because cats treat humans the way kittens treat their mothers, humans who treat their cats like babies simply respond in kind. In fact, some owners believe that any other reaction would ignore the cat's most basic needs.

Highly anthropomorphic relationships with cats also survive because of the cat's solitary nature, which easily reinforces an owner's image of a pampered only child. Dogs are pack or social animals; they normally seek the companionship of others and therefore invite criticism for any abnormal behaviors. So, while dogs who run and hide every time the doorbell rings will receive hoots of derision from friends and neighbors, cats displaying identical body language will usually hear coos of sympathy. We expect dogs to be gregarious and cheerful (that is, like us), but we automatically accept the aloof "scaredy cat" or "fraidy cat" as perfectly normal.

Like deifying our cats, treating them as little fur-suited people generally causes few problems as long as the cat behaves in ways we can comfortably associate with human behavior. However, we must bear

The Human Response: Running the Emotional Gamut

in mind that when we view our cats as people, we essentially use them as mirrors of ourselves. As long as what they do reflects what we find acceptable, it's like looking in a mirror to admire a new hairstyle we find attractive; but if the cat behaves in a way we find undesirable or incomprehensible, it's like discovering an ugly blemish or, worse, a total stranger staring back at us. Whenever owners use human behavior, and particularly their own behavior, as the reference point, "inhuman" conduct on the part of the cat generates every bad feeling imaginable: confusion, frustration, revulsion, guilt, and even self-recrimination. Such feelings strangle any desire to effect positive changes or accept the behavior as normal *for a cat*. When Sharon awakens to find a dead mouse on her pillow, it takes her days to recover from the ordeal because she simply lacks any comprehension of Percy's predatory nature. Consequently if she wants to reconcile her feelings for her pet, she must quite literally start from scratch, working through her emotions and beliefs until she formulates a new definition of Percy that includes this behavior.

Compare this with the effect of a similar offering from Clipper, a Maine coon cat belonging to Charlie and Helen Zubriski. Because the Zubriskis live in a rambling old farmhouse on six wooded acres, their pet's mouse-catching ability constitutes normal feline activity. Charlie often responds to Clipper's lifeless offerings with "Good boy!" and gives the cat a few extra affectionate pats. Although Helen personally finds the predatory behavior repugnant, she recognizes it as normal for Clipper; because of this she simply disposes of the mouse with little, if any, display of emotion.

Anthropomorphic owners also run into difficulties when their cats become ill or die. Painful specters haunt Sharon:

· If Percy can succumb to a urinary tract infection, so could I.

· If Percy fails to respond to medication, that might happen to me.

· If Percy dies from congestive heart failure, I could too.

Such associations may seem farfetched as we sit reading about this phenomenon with our perfectly healthy cats curled up contentedly in our laps. When a beloved pet does become ill, however, anthropomorphic owners often let raw emotion override logic and reason. Every veterinarian has struggled to extract factual, perhaps life-saving information from distraught anthropomorphic owners. Usually these people have become so embroiled in their feelings about how the cat's illness affects them that they contribute to, rather than help solve, any medical problem. Their inability to get past their own emotions leads to feelings of impotence, anger, and guilt, all of which can build to devastating proportions if the sick cat dies.

It may be easy for Sharon to watch Percy leap gracefully from table to bookcase to desk and to compare him to Nureyev or Nijinsky, but it's not so easy to summon up a valid human analogy when Percy licks his scrotum while curled up on her pillow, a scant half inch from her face. And it's hardest of all when he lies sick and dying; whatever sense of balance and perspective may have attended the anthropomorphic relationship, like logic and reason, inevitably vaporizes under pressure.

THE AILUROPHOBIC OWNER

At the other extreme of the human-response spectrum lie the ailurophobes, who, according to *Dorland's Illustrated Medical Dictionary*, don't merely fear cats, they exhibit a *pathological* fear of them. If so, why even bother discussing such feelings in a book dedicated to celebrating the special human/feline bond? As with our discussion of the medieval cat slaughterers, an understanding of such feelings can help us develop useful standards with which we can evaluate our own interactions with our pets.

"Don't be absurd," sniffs Tony Palmieri. "Anyone who hates cats wouldn't own one."

Probably not. But how do you feel when your pussycat barfs up mouse guts on your bed? (This common phenomenon prompted one owner to ask me whether cats are physiologically and behaviorially capable of throwing up anywhere else!) However, when our cats display

The Human Response: Running the Emotional Gamut

behaviors we find repulsive, the line separating ailurophilia from ailurophobia stretches very thin indeed. We already saw what an emotional gamut such owners run, often experiencing anger, frustration, confusion, fear, hatred, guilt, then overindulgence, and, finally, acceptance, before love becomes the dominant emotion again.

Although we might prefer to avoid such situations at any cost, doing so actually makes the problem worse. If we tend to think of ailurophobes as cat haters, then anytime we have adverse feelings toward our cats or their behaviors, we must see this as evidence of our own hateful natures. Instead we should remind ourselves that nine times out of ten, ninety-nine times out of a hundred, our temporary anger and momentary hatred spring from fear, which, in turn, springs from ignorance. By refusing to acknowledge the reality of our feelings, we simply perpetuate the ignorance and do nothing to prevent a recurrence of the negative interaction.

Before we can eliminate ignorance and fear, we must first accept the fact that cats do belong to a different species with habits and behaviors significantly different from our own. Every feline companion conceals a few tricks up its furry sleeve, and while part of the allure of owning a cat comes from these charming or "good" differences, less charming or downright "bad" ones can alarm even the most ardent cat lover—or, perhaps more correctly, *especially* the most ardent cat lover.

The more closely attached to our cats we become, the more likely we will experience episodes of ailurophobia. The more our cats trust and enjoy our company, the more likely they will be to display their full range of behaviors. The wider the range of behaviorial displays we share with our cats, the more likely one or more of those behaviors will upset or repulse us. That such is the case is summed up beautifully in a line from *The Little Prince* by Antoine de Saint-Exupéry: "One always runs the risk of weeping a little if one lets himself be tamed."

Periodic encounters with ailurophobia are the price we pay for a close relationship. Such episodes needn't lurk like skeletons in the closet, however. Rather, by accepting these as a predictable—not evil

—consequence of a relationship with something different, we free ourselves to deal openly and honestly with our feelings.

For example, Tony Palmieri builds his relationship with Lotus on
the rather flimsy illusion that pedigree and price confer some sort of
superiority on his cat. In addition, he also harbors beliefs that her
advantages should enable Lotus to act more "civilized" (that is, more
human) than the ragtag assortment of domestic shorthairs he knew in
his past. Given Tony's previous and highly limiting definitions of an
acceptable range of behaviors, it's inevitable that Lotus will run afoul
of her owner's expectations. First she irritates Tony by excavating her
litter box: "What's she doing in there—building a damn freeway?"
Even more unforgivable, Lotus is a slob. Not only does she scatter
litter during her exuberant forays into the box, she cares little for
personal hygiene. Someone has to wash the expensive pedigreed cat's
face and groom her daily; otherwise her coat becomes caked with food
and tangled with mats. Tony refuses to have anything to do with this
ritual, and the responsibility for keeping Lotus clean falls upon Kim.
Initially this creates no problems because Kim enjoys fussing over Lotus, and Lotus adores the attention. However, during the final week
of her pregnancy, Kim ignores the grooming, and Tony does little
more than feed the cat the week his wife spends in the hospital. Consequently, Kim hardly recognizes the matted furball bounding to greet
her when she returns home with the baby: "Tony, you neglected poor
Lotus!" Tony catches himself thinking about the down jacket he could
have gotten with the hundred and fifty dollars he wasted on this lazy,
irritating cat.

When the Himalayan's natural curiosity draws her to the sleeping
baby's room and Tony catches the cat sleeping practically on top of the
heir apparent, the incongruency between Tony's real and his ideal feline images infuriates him. For months Lotus had whittled away at his
mental image of the delicate, immaculate, and devoted cat who dotes
on him and heeds his every word. When he spies the grimy feline
defiantly ignoring his orders to stay out of the crib, using his son's

The Human Response: Running the Emotional Gamut

handmade quilt as a pillow for her grubby head, something snaps.

Every cat owner I've ever met has had to deal with similar anger and frustration, however fleeting, at some point in his relationship with his pet. Even if Tony manages to control himself sufficiently to just scream at Lotus rather than grab her and shake her roughly before flinging her across the room, he still must confront all his negative feelings. And while none of us finds it easy to acknowledge that any dumb, four-legged, fur-covered being can arouse such passions, this kind of acknowledgment is essential to the formation of a solid bond. If we don't recognize our strong negative as well as our positive feelings as a normal part of a relationship, we deny ourselves the ability to deal with such feelings when they arise. Instead we wind up trying to pretend that those negative thoughts never crossed our minds; or, worse, we feel guilty or disloyal to our pets because they did.

If Tony tries to ignore his ailurophobic reaction, he might create this paradoxical sequence:

Tony tells Lotus to stay out of the baby's crib. He discovers the cat in the crib. An image of grabbing the cat and smacking her crosses his mind. He snatches her out of the crib, shakes her roughly, and tosses her to the floor. Her bewildered and frightened expression cuts through him like a knife. Anger, confusion, guilt, and, finally, exhaustion and remorse wash over him. He seeks out the cat, pets her, and speaks to her soothingly, then gives her a dish of ice cream.

We know that patterns of behavior become incorporated in an individual's or species' repertoire because they serve a purpose: What purpose does this sequence serve? According to Tony, "I want to teach Lotus to stay out of the crib"; but can owner guilt and a food treat (that is, a reward) accomplish that goal?

In many ways, the loving owner who inconsistently and emotionally confronts periodic ailurophobia creates a more trying state for both human and cat than all those cat crucifiers imposed upon felines in the Middle Ages. The medieval ailurophobes hated all cats and everything about them and made no attempts to hide their feelings. Consequently,

THE BODY LANGUAGE AND EMOTION OF CATS

they always responded violently to cats. Once cats learned—by observation or personal experience—to read this human body-language display, they avoided human contact. Compare this with the plight of house cats Lotus and Percy, who may find themselves deified one day and damned the next.

WELCOME TO THE TWILIGHT ZONE

Diluted ailurophobia often takes the form of indifference or emotional neutrality. Although it would seem logical that nonchalance signifies a more humane relationship than violence, indifference can cause problems every bit as severe as physical punishment. Obviously it goes without saying that those who hate cats or feel no attraction toward them shouldn't own them. (Having one "for the kids" doesn't make sense unless the kids can accept full responsibility for the animal's welfare.) But what about the person who says, "I can take 'em or leave 'em"? Leave 'em.

Unfortunately, even loving owners sometimes find that providing a nonresponse or evading a response is preferable to making a negative one. When Tony wants to smack Lotus, suppose he forces himself to turn his back on her and walk away instead. This immediate solution is most certainly preferable to violence, but it serves no beneficial purpose if Tony sees it as his *only* option. This is a subtle but critical difference we need to recognize when we opt for an indifferent or neutral response.

When we choose to ignore a certain feline behavior because recognizing it might cause us (or our cats) pain, we must ask ourselves this question: Am I ignoring this behavior because I fear resorting to a worse display (screaming, violence) or because I believe this is the best response for me and my cat at this time? If we respond indifferently or neutrally because we feel that our desired response is blocked for some reason, all we do is add resentment to that subliminally simmering stew of negative emotions. How does Tony enhance his relationship with Lotus if he subordinates his desire to swat her because he thinks doing so will anger his wife? How does Sharon strengthen her bond with Percy when she buries her feelings that her cat deviously manipulates her? Clearly an unemotional, intellectual approach contributes nothing beneficial to a relationship entangled basically by emotional problems.

On the other hand, if we view neutrality or indifference as a choice, then it becomes a valuable response when we confront aliurophobic situations. If Tony walks (or runs) out of the baby's room because he doesn't *want* to punish Lotus, he sets an entirely different set of options in motion. While it may appear that Tony has ducked the problem, by making a conscious choice to do so he actually assumes control over, and responsibility for, his behavior. Whereas the forced indifferent reaction results in his blaming Lotus for his hostile feelings and feeling victimized, his choice to flee until he can calm down puts him completely in charge of the situation. In such a way Tony can buy time to initiate changes that might improve his relationship with his pet.

THE BONDED APPROACH

Having worked our way from both extremes and surveyed two important intermediate positions, we arrive at the fulcrum on which the best responses pivot. Ideally this should occur halfway between ailurophilia and ailurophobia, and somewhere between anthropomorphism and indifference. Or should it? We began our exploration of typical human responses with a discussion of balance and perspective. Because no real-world relationships ever follow a lock-step set of body language displays and emotional responses, each different display and even the same display at different times possesses unique characteristics as well as basic unifying threads.

Creating and maintaining a bonded relationship with a cat results less from what we do than from what we believe. To bond with another creature we must value the co-creation of the relationship, the choice both have made to share the same environment and exchange body-language displays and responses. While such recognition constitutes the foundation of all bonded relationships, regardless of species, it becomes more crucial than ever when we choose to relate to a cat. Given the cat's solitary nature and our own social orientation, we must forever balance that critical difference in our relationships. When we judge our cat's behavior, we must take into account alien but perfectly normal feline behaviors, such as nocturnal or predatory predilections, before we apply our own human criteria of right or wrong, good or bad, to the cat's actions.

Above all, a bonded relationship with a cat demands that we set aside our "like me," "right," or "wrong" blanket definitions, which may suffice when we evaluate the behaviors of other social species, and embrace and celebrate the differences between human and feline. Although such differences make the bond between human and feline at times a difficult one to forge, the result can be especially powerful.

A bonded relationship with a cat demands a philosophy of paradox, one that replaces the traditional either/or linear thinking with a more multidimensional "combined with" mind-set. On the one hand the cat's asocial nature forces us to accept a certain amount of behavior we find unnatural and alien. On the other, its tendency to treat us like its mother demands that we constantly guard against anthropomorphism lest we deny it its unique feline disposition. The feline senses and habits that equip it so beautifully for a nocturnal existence fascinate and charm daylight-attuned humans, yet we long for our cats to play in our sunlit world and curl up beside us and sleep soundly through the night. We celebrate their independence but chafe at their often obstinate resistance to training and restraint. We adore the velvet softness of their footpads on our arms, but violently pull back from the passive prick of unsheathed claws. Although we delight in the fact that Sylvester will only accept food from members of the immediate family, we deplore his perverse refusal to accept any food except one particular brand and variety.

Inherent in these paradoxes, the very best and the very worst feline behaviors are routinely served up to cat owners to celebrate or condemn on a daily basis. Sometimes the extremes are so obvious that logic screams out that one must surely be "very right" when compared with the other's "very wrong"; and on the next day, the very opposite seems to be equally obvious and true. If we approach our paradoxical cats with an either/or attitude, we inevitably push our cats and ourselves into awkward positions; but if we can think in terms of accepting both extremes as normal under certain circumstances, this combined approach can solve a host of apparent conflicts between us.

Bonded owners don't try to deny or reconcile the paradoxes inherent in their interactions with their pets. Rather they strive for a dual perspective that provides a consistent balance and may encompass the full range of feline body language and human responses to it. From all these ambiguous feline activities and conflicting human responses we can forge an alloy, a synergistic bond based on differences that is both stronger and unlike either of its two components. Bonded owners don't ask, "What did I (or the cat) do wrong?" when things don't work out, but rather "What does this say about my relationship to my cat?" And furthermore, "What do I intend to do about it?" Bonded owners recognize that they are participants in, not victims of, their relationships with their pets.

BONDED PROBLEM SOLVING

Does becoming a bonded owner sound like it requires a tremendous amount of knowledge and skill? In a way it does, but as long as we keep a few basic principles in mind, it's really quite simple. Principle number one: Cats are different, not wrong or right beyond human comprehension, simply different. Number two: Regardless what happens between human and cat, ascribing human definitions of evil, wrong, spiteful, guilty, or stupid serves no useful purpose. Not only do such evaluations do nothing to define or solve the problem, they portray the cat as an unworthy beast and give rise to the obvious questions:

- What kind of person would own such a beast?
- Why waste time correcting any problems that arise with such an undesirable beast?

The answers to both questions could very well undermine a good relationship and destroy an already troubled one.

For example, suppose Sharon defines the problem as being Percy's imperious and selfish demands for fresh fish. Having so categorized her cat, by her own definition she becomes the kind of person who would own a mean, selfish cat. Although individual ideas about people displaying such human qualities may vary, few would perceive them as particularly positive. Every day veterinarians and groomers listen to owners complain about their cats' jealous and spiteful behaviors. Rather than arousing sympathy, such tirades more often make us wonder what's wrong with the *owner* that he or she endures such a relationship.

Even if we share the owner's anthropomorphic opinion that a cat really is spiteful and mean, will altering a particular behavior change anything? Sharon labels Percy mean and spiteful because he makes her drive fifty miles for his fish. Will her opinion of his nature magically alter if she stops fetching the fish? Probably not. If he's mean and spiteful when receiving this exotic fare, surely we can't expect his devious nature to improve when he's denied this treat.

Given the obvious incongruencies of such an approach, doesn't it make more sense to eliminate all negative emotion from the situation and simply state the problem objectively? By doing so, Sharon's previously complex emotional problem becomes: "Percy likes fresh fish, but I don't want to make that weekly trip anymore."

BONDED SOLUTIONS: ACCEPTANCE

Once we eliminate negative emotions and define the problem as objectively as possible, *all* solutions fall into one of four categories. First, we can accept the situation exactly as it exists, including any negative feelings we have about it. Sharon can continue her weekly fish runs and live with the ambivalent feelings they generate; Tony can continue

The Human Response: Running the Emotional Gamut

judging Lotus's behavior and demeanor as beneath her cost and breeding and let these feelings color his relationship with her.

Although accepting these situations may strike you as a poor solution, sometimes acceptance works wonders. Remember Clipper, the Zubriskis' mouser par excellence? Helen cringes at the thought of predatory behavior, but prefers Clipper's hunting to the use of non-species-specific poisons or traps. Consequently, whenever Clipper deposits a dead mouse in the house or yard, she recognizes and accepts her negative feelings about the act, but she doesn't allow them to erode her relationship with Clipper.

Acceptance serves a useful purpose in strongly ailurophilic or ailurophobic situations where we feel unwilling or incapable of making any changes. Obviously, when we feel strongly positive about the cat's behavior and our response to it, why change, particularly if the eccentricity doesn't bother anyone? Recall the cat food commercial in which owner and cat performed the cha-cha as part of the feeding ritual. While noncat people may have found this interaction bizarre, the success of this particular ad campaign tells us that many cat owners found this behavior quite acceptable.

However, as in the case of the little girl with the little curl in the middle of her forehead, the very, very good always harbors the potential to be very bad as well. Owners who accept extreme responses, no matter how positive, must live with the reality that any negative feline display—be it misbehavior or illness—could precipitate equally powerful emotions at the opposite end of the spectrum. As long as we understand that this potential exists and can accept both extremes as normal, the relationship will endure. Unfortunately, extreme responses often arise exactly because owners lack the confidence necessary to create a more balanced relationship. Accepting a cat's behavior and our responses as unchangeable and beyond our control frequently leads to a relationship in which the negative interactions escalate over time until they reach explosive levels. At that point Sharon may abandon Percy, or Tony may slam Lotus against the wall.

THE BODY LANGUAGE AND EMOTION OF CATS

CHANGING FEELINGS FOR THE BETTER

Our second option involves accepting the cat's behavior but changing our feelings about it. This simple solution provides a powerful tool for building a solid relationship with the cat—more so than with any other species. The reason more owners don't take advantage of this option rests squarely on its overriding demand: knowledge. Anyone can cheerfully accept normal behavior. But if we don't know what's normal or if we equate different with wrong, the most we can expect, barring changes in the cat, is acceptance accompanied by a raft of negative feelings. Although we may achieve moderate success using this approach with other social creatures, whose behavior parallels our own, it fails miserably for cat owners. If we equate different with abnormal and wrong, we become trapped by our own definitions into accepting not only "wrong" behavior in our cats but our equally unpleasant responses to them. To extricate ourselves from this nonproductive process we must learn as much as we can about normal cat behavior, thereby establishing a standard by which we can evaluate our cat's behavior in terms of its uniqueness.

Let's see how this works. The major detrimental effects on Sharon's and Percy's relationship arise from Sharon's feelings of manipulation when she makes the weekly fish run under what she considers less than ideal circumstances. Suppose Sharon decides she doesn't like these negative feelings and wants to eliminate them from the relationship. First she must determine whether withdrawing the fish from Percy's diet will harm the cat. She takes Percy in for a thorough examination, after which the veterinarian assures her that her cat not only can forgo the fish, but should. The steady fish diet, it turns out, could create problems for him as Percy grows older. Armed with this knowledge, Sharon gains the confidence to deal with any guilt or other negative emotions she might suffer or perceive Percy to suffer as she implements his new diet. Depending on her own and Percy's personality, she may simply stop the fish cold turkey, or she may find it easier for both of them if she gradually diminishes it over a period of two or three weeks. If the veterinarian suggests that she continue the finny supplement because of some rare digestive deficiency, Sharon can continue making her weekly trips fortified by the knowledge that she's doing something beneficial for her pet and not responding to his exorbitant and tyrannical demands.

Regardless of the outcome, increasing her objective knowledge about Percy's normal needs enables her to take control of the situation rather than allowing it to control her. Even if she chooses to continue the trips, her negative feelings about them will diminish once she recognizes that this practice reflects a personal choice.

Tony Palmieri experiences equally positive results when he seeks more information about Himalayans in general and Lotus in particular. In fact, an afternoon with the breeder changes Tony's entire view of his pet. For one thing, the breeder informs him that she's emphasized temperament in her breeding programs because she feels that owners, particularly those with young children, want a pet they can trust. Tony vigorously shakes his head in agreement. Unfortunately, animals displaying this social characteristic seem to do so at the expense of the more solitary survival skill of grooming. "Between the breed's long coat and my line's preference to be with people more than groom, the price you pay for temperament is matted fur," notes the breeder. "Still, because so many of my kittens go to families, I think it's worth any inconvenience. These cats should be groomed anyhow, and because they're so people-oriented they love having it done."

Following his visit to the breeder, Tony sees Lotus in a completely different light. As he gazes at the cat, contentedly purring at Kim's side as his wife nurses the baby, he realizes that what's normal for Lotus—her marvelous temperament—more than erases any negative feelings he may have about her sloppy personal habits.

"Hey, cat, c'mon over here and let's get you cleaned up." Lotus's delighted rubbings and purrings immediately expand Tony's new positive feelings about this cat he once disdained.

THE BODY LANGUAGE AND EMOTION OF CATS

CHANGING BEHAVIORS: TEACHING OLD CATS AND OWNERS NEW TRICKS

While we can indeed change a cat's behavior, we can't do it the way we can train a dog. To begin with, cats aren't dogs; not only do they exhibit some very different normal patterns of behavior, they also relate to people quite differently. Therefore, this option requires not only knowledge but also the skill to implement that knowledge in a way that will guarantee the desired results.

For example, suppose Helen Zubriski decides she wants to stop Clipper's hunting forays. A typical dog-training technique would involve the application of a negative stimulus every time the behavior occurs. However, because Helen despises the behavior, she refuses to stay up nights stalking Clipper so that she can yell at him or blast him with a water pistol every time he zeros in on a mouse. More likely, any attempts she makes to change his behavior will·link her punishment to the already dead mice Clipper leaves lying about or brings to her. If Helen makes enough of a fuss long enough, she may eventually convince Clipper not to deposit his kill in particular places. Although this does nothing to halt the predatory behavior, it removes the evidence from Helen's view. Out of sight, out of mind: If dead mice no longer litter the landscape, Helen can now choose to believe she "trained" Clipper not to kill them.

This isn't to say that cats can't be trained, but rather that we must get to know our cats very well before we attempt to change them in this manner. Often such knowledge leads us to change ourselves or the environment in order to achieve the desired results more effectively. For example, we know that cats rely more on their sense of smell when evaluating immobile objects. Consequently, the sleeping owner (or baby) gets sniffed more often than the awake or moving one. We can see how the normally nocturnal cat, perhaps bored with the lack of activity in the house or aroused by something going on outdoors, seeks out its owners. Finding them fast asleep, it switches to olfactory perception, zeroing in on areas of greatest scent, that is, those with the highest

The Human Response: Running the Emotional Gamut

moisture content. Because the mouth and nose provide the most readily available source of scent data, the cat concentrates its efforts on these areas.

An awakened owner often attaches symbolic meaning to the cat's simple fact-finding display:

"Are you hungry, Clipper?"
"Do you want to go out?"

If owners then respond by getting up and feeding the cat or letting it out, they actually teach the cat to link its normal unemotional and nonsymbolic sniffing behavior to food or going out. Consequently, when owners tire of such behavior or feel manipulated by it, it makes more sense for them to change their own behavior, that is, stop responding to the cat, and break the symbolic connection. This approach gains added value because it doesn't necessitate the use of punishment so often associated with changing or discouraging perceived negative displays. In general, cats don't respond gracefully to punishment, perhaps because their natural response to people as kittens to their queens leads them to accept a certain amount and/or intensity of punishment as normal. When that amount or intensity is exceeded, it becomes threatening and alienating, causing the cat to freeze, fight, or run away —none of which succeeds in resolving the negative behavior without undermining the relationship.

We may loosely compare this method to that used by parents who nag or punish their children to stop certain behaviors "for your own good." Although the parental reasons may be theoretically sound— don't drink, don't do drugs, sit up straight—the child often resists the parents' approach to eliciting change as much as or more than the change itself. Consequently, the parents now face two problems: the initial negative behavior and the child's rebellion.

Asocial cats may also view punishment differently from humans, dogs, and other social species, who use it primarily as a mechanism for maintaining pack structure. Because such pack structure lies outside normal adult cat behavior, punishment may serve as a destabilizing influence rather than the stabilizing one it is among social groups. Rather than triggering the "change" or "do something different" reaction common among social animals, who will react instinctively to pacify a more dominant pack member, cats may very well perceive any unexplainable discomfort or surprise only as a threat to their survival. As a result, fear causes them to freeze, fight, or flee. If an owner wishes to stop a cat from doing something (scratching the furniture, for example), punishment will *appear* to work because any one of the three fear displays preclude clawing; on the other hand, the owner runs the risk of converting a clawing problem into a biting or scratching one if the cat opts to fight rather than freeze or flee. If Clipper opts to jump off the bed when Helen takes a swing at him, she can say that her punishment or threat of it solved the problem. However, if Clipper sees her swing as sufficiently nonthreatening, he may see this human display as an invitation to play and gleefully take a few swipes at Helen. If he considers her raised arm threatening but surmountable, he may attack. Regardless of which approach she takes, Clipper responds to Helen's behavior and not to the fact that she didn't like him sniffing her. From this we can see that successful cat "training" depends much more on the owner's interpretations of the cat's behavior than on the behavior itself.

CHANGING THE ENVIRONMENT

Although environmental changes can solve many of the problems that plague human/feline relationships, owners who lack confidence often see these as giving in to the cat. For six months I endured Maggie standing on my books to look out the window beside my desk. Because those books were the smallest and lightest of the collection spanning the length of my desk, her weight easily upset the delicate balance that kept the entire arrangement together. I can't recall how many times I yelled at her as I retrieved book-end and books from the floor. Nor can I remember how many times I dropped my pen midword to hiss profanely and lunge at her as she sauntered toward the window.

"Why don't you just move the books?" asked my husband after a particularly athletic lunge resulted in my scattering books and papers from one end of the room to the other.

"Hah, easy for you to say!" I snorted, astounded at his naïveté. But his look of total incomprehension, coupled with Maggie's equally questioning stare, soon eroded my meager defenses, all of which derived from pure emotion and no logic whatsoever. I moved the books that same day, and now Maggie routinely visits me on her way to and from the window. What used to be a negative interaction perpetuated because I insisted on trying to change her "as a matter of principle" has now become one of my favorite shared experiences, thanks to a minor environmental change. Once again I was reminded how easily we can lose sight of the overall relationship by hanging on to incongruent beliefs.

THE FINAL OPTION

Finally, we must consider an option most of us would rather ignore: termination. Although many recoil at this option, just considering it can be most empowering. So many times owners confronting behavioral or medical problems try to pretend that this option doesn't exist rather than dealing with it openly and honestly. In such a way they often prolong a problem unnecessarily, in terms of both time and negative emotional involvement.

For example, suppose the Palmieris decide that between their work and the new baby, they simply can't afford the time to groom Lotus properly. While perhaps few see grooming as a legitimate reason to get rid of a cat or have it euthanized, if Tony and Kim honestly consider these options, their relationship with Lotus immediately comes to the fore. If they can say, "Good heavens, we'd *never* think of getting rid of Lotus, no matter how ratty she looks!" they free themselves to concentrate all their efforts on finding an alternative solution, such as having her groomed professionally, hiring the teenager next-door to do the job, or shaving Lotus down to minimize any grooming needs. If the idea of getting rid of Lotus even remotely appeals to them, the Palmieris are then forced to confess the real problem—their relationship with Lotus—rather than its minor manifestation, the grooming dilemma. In a relationship lacking deep commitment, one problem will simply give way to another, with owners and cat soon swept up in a vicious cycle of stress-related medical and behavioral problems. Can anyone, regardless of how frightening or repulsive they find the idea of giving a cat away or having it euthanized, condone the quality of such a life?

The termination option carries special significance when applied to cats. Given their asocial natures, it would appear that any long-term commitment a cat makes to any relationship would result from choice rather than genetics. We can't overlook the possibility that problems in a relationship might stimulate cats to choose termination themselves. Given the differences between a social human and an asocial cat, we could be tempted, from a purely unemotional behaviorist's point of view, to say that in a relationship between human and feline, we need the cat more than the cat needs us.

While emotionally our egos may find that possibility hard to accept, many of us do cherish beliefs about domestic dog versus domestic cat survival which support this contention. Many of us believe cats are much more able to fend for themselves than dogs. For a pup to survive in the wild we think it need find, or be found by, people; few of us really believe that a Chihuahua or miniature poodle pup could make it on its own. On the other hand, we expect cats to summon up latent hunting and other survival skills that would enable them to adapt easily to a new environment.

This line of thinking sends confidence-lacking ailurophiles scurrying to offer their pets all sorts of special inducements and privileges to tempt them to stay, while their anthropomorphic friends intensify their efforts to convert Puff into a dependent baby as rapidly as possible. Those leaning more toward the ailurophobic end of the scale may see just one more reason to distrust cats: "Sure you *act* like you love me, but if someone better came along, you'd leave in a flash."

However, for the confident, knowledgeable owner any fears regarding the cat's choice to leave pale beside the joy of knowing that it chooses to be with them. Too often we overlook this unique aspect of feline interaction or choose only to see its more "positive" manifestations. People frequently talk about cats entering their lives at significant times. While many routinely get a cat or kitten to mark events such as the birth of a baby or moving into a new home, or to partially fill the void associated with the loss of a loved one, it's amazing how many strays or unwanted kittens simply materialize on one's doorstep when such events occur; even high-rise-apartment dwellers have experienced this phenomenon.

The introduction of cats into households under such circumstances always carries a strong emotional charge, but their disappearance under similar inexplicable circumstances often shocks us even more. If we perceive the cat's choice to enter our lives as a good omen and a positive reflection on us, we condition ourselves to accept the converse: The disappearance of that cat forbodes bad and indicates some personal unworthiness. Like being in love or mastering the art of using chopsticks, this uniquely feline here-today-gone-tomorrow syndrome is readily recognizable if one has experienced it, but beyond comprehension if one has not. It has happened to me three times, and all three cats were extraordinary creatures who instantly captured my heart and filled a particular void in my life at those times. When each just as suddenly disappeared, like all owners I first feared that the cat lay ill or injured somewhere, so I searched high and low, all the while feeling more and more inadequate. As time went on, I clung to the idea that the cat had died because I couldn't bring myself to consider that it might have chosen to leave. Again, like many owners whose cats magically appeared, I believed that all three were in such poor shape when they arrived that they would have died had I not intervened. Therefore, the grief I felt over their possible demise was tempered by my belief that

THE BODY LANGUAGE AND EMOTION OF CATS

I had postponed the event and given them a "good life" with my care.

Why did the logical explanation escape me? If they simply wandered into my life one day, what's to say that they wouldn't wander out and enter a relationship with another person another day? Surely every veterinarian has confronted the dilemma of examining a "stray," obviously well loved and contented with its new owners, while wrestling with the nagging feeling he or she has seen that cat somewhere before—but with someone else. When a friend moved into a new neighborhood with her four sons, the cat living two houses down the block suddenly decided that hers was a much more suitable household. The owner initially tried to lure the cat back with food treats, but soon realized that the cat preferred its new home. In this particular case, the previous owner's confidence in herself and her cat allowed her to establish a completely different kind of relationship with her pet without experiencing any negative feelings. Other times the new owners may be branded as catnappers or worse by those who can't understand how a cat could do such a horrible thing.

When we study territoriality, we'll also discover that some cats attach themselves to property whereas others attach themselves to people. The former often wind up getting sold with the house; even if the owners do try to take these cats along, they keep returning to the old homestead. Other cats will live anywhere with their owners, but nowhere with anyone else. If such specificity is known to exist, isn't it possible that some cats are attracted to specific kinds of people or to those having specific needs? We all know people who love their work ministering to people in their hour of physical or spiritual need and others who adore children or pets. None of these people can imagine a quality life that doesn't include such interactions. Is it possible that there are cats who are attracted to people in need as well as those who prefer to be around children, or even dogs or horses? If such is the case, when the person resolves the crisis, the kids grow up and leave home, or the dog dies, does it seem so impossible to imagine these cats seeking out other households where these preferred conditions exist?

The Human Response: Running the Emotional Gamut

Most cat people would undoubtedly shout, "Yes!" However, even in this age of technological and scientific answers for every question even noncat people would take note of a strange black cat crossing their paths, particularly on Friday the 13th. So widespread is the link between black cats and bad luck that even the most objective noncat people would be a bit unnerved if that black cat persisted in following them and perched on their doorstep. If we acknowledge (however grudgingly) the relationship between the appearance or disappearance of cats and "bad" events, why not consider this similar connection between people and "good" events? While behaviorists of a more objective mind may cringe at such suggestions, totally ignoring or discrediting the large amount of circumstantial and anecdotal evidence supporting them, such links would seem to deny the possibility of experiencing yet another unique quality of the human/feline bond.

Regardless of why or how termination occurs, the majority of the negative responses associated with this option come from our unwillingness to face such a possibility squarely and honestly. It's our willingness to weigh the option when problems arise that not only gives us the strength to make it through those troubled times but also gives us the peace and confidence fully to enjoy every other aspect of our relationship.

Now that we realize what remarkably variable and adaptive companions we humans make for our paradoxical cats, let's see how our new understanding can equip us to meet the challenges created when nocturnal and diurnal species meet. Is a uniform dull gray our only common ground? Must one or both give up a deeply rooted environmental orientation? Will Fern and the Cruikshanks engage in nightly psychological warfare for as long as they remain together?

5

NOCTURNAL BEHAVIOR: WHAT A DIFFERENCE A DAY MAKES

*I*T had been a long day for Ben and Gerri Cruikshank. The move into their new home had taken the starch out of them as periodic thunderstorms and intervals of jungle-like steaminess, not to mention clouds of mosquitoes, descended on the weary movers. With most of the furniture finally in place, Ben and Gerri shared a long, cool shower before crawling between crisp, clean sheets for a sorely needed night of sleep.

Thump! Bang! "Oh no, not again," mutters Gerri, barely daring to breathe lest she call attention to herself. "I thought Fern was asleep." In fact, Fern the Cruikshanks' long-haired calico *had* slept most of the day.

Rattle, clang. Snap! "If I weren't so tired," groans Ben, "I'd get up and strangle that cat." Gerri knows her husband would never hurt Fern, any more than she would; but the thought of another lost night of sleep makes her want to cry.

A somewhat different nightly ritual occurs at the Nortons' house on the other side of town. After watching the eleven o'clock news Karen Norton retires upstairs to brush her teeth and prepare for bed. Meanwhile her husband, Bruce, commences to hunt for the family cat, Marmalade, an orange tiger shorthair whose name, according to his owners, reflects "his sweet-as-jelly personality as much as his color."

Although "putting out the cat" is a time-honored tradition in many

134

Nocturnal Behavior: What a Difference a Day Makes

households, for the Nortons it's turned into a nightly battle of wits. Not only doesn't Marmalade want to spend the night outdoors, he wants to spend it snoozing in the Nortons' bed, something Bruce finds intolerable. When they first got the cat, Marmalade slept contentedly right next to his owners while they watched the evening news, so putting him out afterward posed no problem except for some minor feline resistance. However, when Marmalade reached nine months of age, Bruce and Karen had him neutered, and from that day on, getting him to go outdoors became a struggle.

Take tonight, for example. First Bruce shuts both doors to the living room, then he carefully searches all of Marmalade's favorite hiding places; not finding him in that room, Bruce blocks all escape routes from the dining room and repeats the process. Meanwhile he hears Karen doing the same thing in the children's rooms upstairs. Finally Bruce works his way into the kitchen. He stands stock-still, hoping to pick up a telltale sound from the fugitive cat. A slightly ajar cabinet door catches his eye. He creeps toward it stealthfully, then lunges. Marmalade explodes through the containers of dish detergent, cleanser, scouring pads, and furniture polish like a furry orange cannonball.

"Strike!" crows Bruce as every container spills across the kitchen floor. With patience ebbing and anger flowing, Bruce pursues the fleeing cat into the dining room. Upstairs Karen stares unhappily at the bedroom clock and hopes Bruce won't wake the children—again. It's almost midnight when Bruce finally comes to bed.

As we begin our discussion of specific feline behaviors that exert the most influence on the human/feline relationship, we must first confront some paradoxes inherent in the study of animal behavior itself. Animal behavior, or ethology, is a science and, as such, adheres to the rules of science. Although the practice of science encompasses many rules and regulations, the two of most concern for cat owners are the elimination of variables and the quest for reproduceable results. Any well-designed experiment in animal behavior strives to control the animal's environment as much as possible in order to eliminate variables so that other researchers may repeat the experiment and obtain results that support or negate the original findings. Obviously, in order to provide verifiable new facts about animal behavior, the behaviorist needs to set up his or her experiment in such a way that any results of others reproduce and thus reinforce the original conclusions.

Sensible? Sure, but a quick glance around your own home most likely reveals an environment significantly different from the one the Nortons share with Marmalade or the Cruikshanks share with Fern. And all three of these differ significantly from that of a research facility in Philadelphia or Chicago. Therefore, although controlled experiments can give us new insights into feline behavior, we must always bear in mind that no controlled experiment necessarily matches the particular environment or set of conditions operating in our own homes.

Similarly, we must also put the data gleaned from the study of wild felines into proper perspective. While domestic and wild cats do exhibit many similar behaviors, we must guard against comparing the incomparable. For example, of all the different species of wild felines available for comparison to the domestic cat, the lion is viewed by many people as the wild counterpart of *Felis domestica*. Ironically, because lions are social rather than asocial or solitary creatures, they tend to behave less catlike and more human- or dog-like than any other felines. Consequently, we probably compare our cats to lions because the latter remind us more of ourselves than of our feline pets.

Finally we need to distinguish between wild and feral cats. Wild cats are nondomesticated species; feral cats are domestic cats who have reverted to a wild(er) state. Although all felines, regardless of species, do display some common behaviors, what holds true for a tiger might not necessarily hold true for Marmalade and Fern, and vice versa. However, the study of feral-cat populations often enables us to bridge the behavioral gap between the very wild and the very tame, showing us how cats behave in the presence of humans on whom they do not depend for survival.

Nocturnal Behavior: What a Difference a Day Makes

CIRCADIAN RHYTHM: GETTING INTO THE SWING OF THINGS

Perhaps one of the most exciting areas to emerge in biology and medicine over the past several decades is the study of biological rhythms. We, along with the writer of Ecclesiastes and the poets and songwriters who parody him, intuitively recognize that

> To every thing there is a season
> And a time to every purpose under the heaven
> A time to be born, and a time to die
> A time to plant, and a time to reap.

Furthermore, we recognize that natural rhythms don't confine themselves to the seasons but permeate all levels of existence. Not only do various living organisms engage in greater or lesser activity as summer gives way to fall or fall to winter, but they also vary each individual day, with periods of activity alternating with periods of rest. Compare the image of a serene winter's evening and a sleeping cat curled up on the hearth with that of a spring morning and that same cat joyfully leaping at butterflies.

One of the most charming examples of cyclicity comes from eighteenth-century naturalist Carolus Linnaeus. Noticing that flowers of the same variety open or close at the same time each day and that other varieties obey equally predictable timetables, Linnaeus created the first "flower clock." By planting flowers that opened or closed at progressive hours in a circle forming a clock face, on a sunny day he could tell the time within half an hour. If the African marigold was opening, he knew it was around 7:00 A.M.; and when the evening primrose opened, at 6:00 P.M., he closed up his books for the day and turned toward the evening's festivities.

Linnaeus's flower clock and other early experiments supported the idea that circadian (literally, "around or about a day") rhythm is not only the shortest cycle living things experience, it is also inextricably tied to light. Time and further research, however, have proven that living organisms resist easy classification and rigid control when it comes to their cycles. In fact, what makes us human and our cats feline involves not so much a precise collection of measurable rhythms and activities as an intricate orchestration of intermeshing physiological and behavioral patterns, each following its own peculiar rhythm. When we say that humans are diurnal or daylight creatures, this doesn't merely mean that you and I prefer to work and play when the sun shines; it means that countless cycles work together within us to create a sense of well-being that makes us more alert during daylight hours. These cycles dictate all sorts of responses:

- When and how frequently we urinate (usually more in the morning)
- Our hormonal levels
- Our sensory responses (usually more sensitive in the late afternoon)
- Our body temperatures (usually lower in the morning)
- The numbers and kinds of cells in our blood

Similarly, our cats obey physiological and behavioral cycles that ensure their well-being in their particular environment. Because Fern's barn-cat mother is an avid hunter, her physiological and behavioral cycles support greater voluntary-muscle activity during limited-light periods and relegate more passive functions, such as nursing her young and grooming, to the nonhunting hours of the day. Consequently the definition of a normal or healthy individual includes not only what a particular system does but also *when* it does it. Think of the body as a full orchestra playing Tchaikovsky's *1812 Overture:* If those in charge of the chimes and cannonfire don't synchronize their parts with the entire orchestra, the results appall rather than please. The responses associated with hunting or nursing each play vital roles in orchestrating a cat's life; a queen whose milk began flowing in the middle of a hunt or who experienced an urge to stalk or kill while tending to her young would experience behavioral and physiological cacophony.

Nocturnal Behavior: What a Difference a Day Makes

A TIME FOR EVERY PURPOSE

Once we recognize the marvelously intricate balance that orchestrates life, the fact that the nocturnal feline meshes at all with our diurnal human habits borders on the astonishing. Consider the life-style of another nocturnal animal, the flying squirrel, a creature more common than the gray squirrel even in areas as populated as metropolitan Washington, D.C. Yet few city or urban dwellers know about flying squirrels, because their nighttime habits place them in a completely different world. I didn't even know they lived in my attic until I left the door open and Maggie went on a hunting spree one night.

From this simple example, we can easily appreciate one purpose of nocturnal behavior: environmental economy. Species practicing complementary daytime and nighttime habits can occupy the same environment without getting in each other's way. The rising and setting of sun and moon, the appearance of certain species and the disappearance of others, and innumerable individual cyclic changes all contribute to the flying squirrel's forays into the night and its retreat to its nest at dawn. The sun comes up, birds begin to sing, bees and dragonflies zip about, hawks float soundlessly overhead, tree squirrels and chipmunks scamper over the stone walls. When the sun goes down, owls, moths, flying squirrels, and mice fill the niches vacated by the others. In such a way, the same limited environment fulfills the needs of both diurnal and nocturnal species.

Not only does this system increase the environment's capacity to support a wide variety of different species, it also enables the mechanism to function continuously. Because nature, like the individual, consists of a multitude of cycles weaving around and through each other synchronously, it's easier to keep the system running than to start and stop it periodically.

Having two differently circadian orientations—one nocturnal and one diurnal—also enables those within the respective groups to develop physiological and behavioral characteristics appropriate to their niches. For example, even though we know it's possible for humans to func-

THE BODY LANGUAGE AND EMOTION OF CATS

tion at night, we also know that such activity must override certain inherent diurnal cycles, an accomplishment more successfully achieved by some than others. Many entertainers consider themselves "night people" and adapt well to such a life-style; I personally find an early bedtime and awakening between 4:00 and 5:00 A.M. suits my needs the best.

While individuals may vary in their cyclic preferences, numerous studies have shown that the worst of all possible working worlds involves the "swing shift." When people work 8:00 A.M. to 4:00 P.M. one week and midnight to 8:00 A.M. the next, their bodies never have sufficient time to adjust and coordinate all the various cycles necessary for maximum functioning. Consequently, although a person may feel better working one shift than another, the unnaturalness of the change coupled with the short duration of each shift guarantees that he or she will never be physiologically or psychologically in harmony. Because wild animals dwell in one state or the other, they not only avoid the trauma of such shifts, they gear their entire physiological and behavioral mechanisms to enhance their survival and performance in the preferred time frame. Although researchers seldom use human data to predict possible domestic-animal behavior, pets can suffer from the effects of swing shifts, too. The Cruikshanks never experienced any problems with Fern until the move altered her schedule. The Nortons' and Marmalade's problems arise less from a specific negative feline behavior than from the owners' erratic attempts to align two conflicting ideas regarding "normal" feline cycles.

THE FELINE INTERLOPER

Does *Felis domestica* interrupt the rhythm and beauty of this natural music with all the cacophony of a kitten bounding across a keyboard? You'd think so, but in fact most of the time it doesn't. All the research regarding what happens when a diurnal species turns into a nocturnal one or vice versa might indicate that our cats should be physiological and psychological wrecks, but for the most part, they appear quite

Nocturnal Behavior: What a Difference a Day Makes

confident and in control of their lives as a species. On the other hand, can we explain the cat's susceptibility to a whole raft of viruses, some of them fatal, as a result of the stress that attends reconciling nocturnal tendencies in diurnal households? Is it possible that this shift has rendered cats more vulnerable to disease in daylight? Have we inadvertently doomed *Felis domestica* to be an out-of-time traveler in order to enjoy its company?

Of course such thought-provoking questions can stimulate guilt: What kind of a demented person would force an animal to ignore its most intimate and critical physiological and behavioral patterns just so that he or she could pet its soft fur and hear its melodious purr? However, guilt serves no more purpose than ignoring the implications a shift from a nocturnal to diurnal life-style may have for a particular individual.

Unfortunately much of the scientific data regarding feline cyclicity seems to cloud rather than clarify the issue. Although many behaviorists acknowledge the cat as a nocturnal being, surprisingly little data actually confirm this premise. Furthermore, even though many researchers in one area of science ascribe nocturnal behavior and all its supporting physiological underpinnings to the cat, scientists in other areas treat cats as if their cyclicity corresponded to our own. Much of the research and observations regarding biological rhythms and sleep patterns have involved rats and cats, both nocturnal species, which would seem to preclude the use of this data when predicting diurnal human behavior. When I asked one researcher why he studied these particular animals, his response was quite revealing: "Rats are cheap, and I like cats." As I observed the obvious affection between this scientist and his test animals, I couldn't help but recall the Guthrie/Horton and Moore/Stuttard experiments with the human-pleasing door-opening cats.

The more I read about the feline circadian rhythm, the more confused I became. To be sure, the feline predilection for rodents makes a strong case for its nocturnal nature. However, its strong bond with the human species would seem to contradict a nocturnal definition. This line of reasoning led me to question whether the domestic cat is currently, and perhaps precariously, balanced on the evolutionary fence, neither completely nocturnal nor diurnal. And this question suggests the paradox: Is it possible that rather than being neither, the cat is, in fact, both?

In order to come to grips with this paradox, we must delve deeper into the external and internal factors that play a role in nocturnal displays.

MESSAGES FROM THE ENVIRONMENT

Recognizing that individual feline behavior results from the harmonious intertwining of numerous biological cycles rather than simple on-off states, we can more easily imagine that the light orientation may vary a great deal within a species. While some species and individuals may be truly nocturnal and others fully diurnal, still others may be most active during the twilight hours of dawn and dusk. More aloof and independent Fern may prefer to patrol when minimal light and few humans prevail. Gregarious, people-loving Marmalade may appear to set his biological and behaviorial clock to the eleven o'clock news, just like his owners.

In general, wild cats migrate toward a particular cyclic orientation for one of two reasons:
- It decreases the probability they will get eaten.
- It increases the probability of their catching something to eat or finding a mate.

A fine, but troubling, example of the first comes to light when we study the shifting activity patterns of lions in Kruger National Park in South Africa. When people began hunting lions, the naturally diurnal lions responded by shifting their periods of activity toward darker hours, a phenomenon displayed by many diurnal species who have fallen prey to man.

This mechanism also functions in reverse: The habits of a predator's prey plays a crucial role in the former's life-style. Consequently, one way to determine what's normal for a cat is to consider what it eats. A cat free to develop a daily cycle that will put the most prey on its table with the least expenditure of energy will synchronize its activity with that of the rodent population. Rats and many varieties of mice experience two peaks of activity: In general they move around more at night, but many prefer dawn and dusk as their periods of maximum activity. Other small rodents exhibit short bursts of activity throughout the entire twenty-four-hour cycle. Similarly, if we feed our cats once or twice a day, we can expect the cat to be more active during those two periods. And these cats' activity periods may differ from those of a house cat whose bowl of dry food is available at all times.

The sun also affects the daily patterns of wild feline activity. As the days lengthen and light intensity increases, nocturnal animals delay their periods of activity. Consequently mice scamper around for longer periods during winter nights than during summer ones. However, their winter activities begin later after sunset and end earlier before sunrise. Therefore a cat hunting such prey could conceivably alter its periods of rest and activity throughout the year, experiencing a longer, more leisurely hunting interval in the summer, and a shorter more intense one in the winter. This explains why indoor/outdoor cats often want to go out earlier in the evening and come in later in the morning in the summer. While owners who rely on their pets for rodent control might accuse their cats of slacking off during the winter, this may not be the case at all. Their cats, like the mice, may simply be compressing their activity into a shorter time span.

From this we can see that biological and behavioral rhythms are simultaneously remarkably stable and flexible. Thus a young kitten in a laboratory, cut off from all environmental cues, will still establish a pattern of activity, rest, and sleep that consistently adheres to about a twenty-four-hour day. On the other hand, within that twenty-four-hour cycle individuals will readily rearrange their schedules to fulfill their needs. Because nature functions on a more flexible biological clock, in contrast to the rigid one of the civilized day, such variations offer a real survival advantage. Imagine all the wasted energy and effort if cats expected mice to appear to 6:00 A.M. and 6:00 P.M. rather than dawn or dusk.

The activity patterns of a successful hunter in one geographical area may differ markedly from those of a member of the same species somewhere else. Furthermore, even the activity of predators hunting the same species may differ in a given area, depending on the time and exact location. Back in the Middle Ages, when humans despised both cats and rats, both species may have been much more nocturnal and secretive in their habits than in ancient Egypt, where a more tolerant view prevailed. And neither of these orientations may parallel the ones preferred by those same species living in a San Francisco suburb today.

Light also contributes to an animal's reproductive success within its environment. In this case, it appears that the rate of change exerts more influence than the actual amount of light. Consequently, in temperate climates, where the amount of daylight increases or decreases most rapidly in March and September, these changes trigger the major breeding seasons, with the intensity of both seasons related to the presence of a reliable food source to support the young. Conversely, this also explains why some of the more highly bred people-oriented feline breeds may exhibit breeding cycles apparently unrelated to light or seasons. Because people function as the food sources, such altered cycles reflect the cat's relationship to human rather than to mouse or other prey species.

MAKING LIGHT OF THE NIGHT

Specific adaptations of the feline sense must efficiently equip the cat to meet two different criteria for night life. First, these senses must function in a limited-light environment; second, they must facilitate survival in that environment. Think about your own response to limited light. Although I can make my way around my yard at dusk and even at night, I can hardly locate a mouse under these conditions. On the other hand, Maggie not only moves quite freely in this environment, she can locate specific objects with deadly accuracy.

We already noted that the eyes of night-active animals exhibit anatomical and physiological differences from those of light-active species. However, within these nocturnal species the eyes of predator and prey each develop to best serve their different needs. Because the feline predator's food can move, the cat needs a more refined sense of vision than a mouse, who needs only to locate an immobile ear of corn or pile of grain.

To be sure, none of the cat's physiological and anatomical ocular variations enables the cat to see as well at midnight as you and I can see at noon. On the other hand, the cat doesn't need that kind of vision, nor is it dependent on vision as its primary sense. To get a better understanding of how these variations suit the cat's needs, let's accompany Fern as she settles into the Cruikshanks' new home.

Ben and Gerri got Fern from Ben's uncle when she was about ten weeks old. Ben's uncle was a dairy farmer, and Fern came from a litter of four, born in his barn to a shorthaired calico known only as Barn Cat. Fern's already long fur, delicate features, and seemingly cheerful acceptance of harsh farm life immediately endeared her to the suburbanite Cruikshanks.

The night of the big move to the new house, Fern focused on only one thing—exploring her new home. For almost a week her life had been chaotic. Not only wasn't she allowed outdoors at the old house lest she get lost or frightened by the movers, her owners locked her up in various rooms for safekeeping during the move itself. Although Ben and Gerri lavished affection on her the instant they relaxed after each long day of packing, cleaning, and moving, every evening Fern confronted a new crop of boxes as well as the disappearance of familiar chairs, lamps, bookcases, and other objects.

Any individual lacking detail vision needs to know its territory. Remember playing "Blind Man's Bluff" as a child? Your ability to catch the other person—your "prey"—directly depended on your knowledge of the environment and its stability. If you played the game in your own backyard, where you knew every stump and fence post, you moved much more confidently than if the game took place in a new playmate's garage. Because you knew your yard, not only didn't you have to see it, you could concentrate all your efforts on locating your target with your other senses.

Because Fern had spent sufficient time with her semi-"wild" mother, her nocturnal, territorial, and predatory instincts had been reinforced much more strongly than those of Marmalade, whose mother was the highly dependent house pet of a most solicitous owner. Consequently, as darkness and quiet slip across the Cruikshanks' household, Fern sets about exploring. Because her physiology gears her to be a predator, her eyes are relatively close together; although this decreases her visual field considerably, it does increase her depth perception, a critical skill for a hunter who must accurately pounce on prey. However, Fern gains this advantage at the expense of two other visual capabilities most rodents possess. Not only can most rodents see more than twice as much to the sides as cats, their only blind spot includes a relatively small area behind their heads. Compare this with the limited peripheral vision of the cat, who can't see what's happening in more than a quarter of its total visual field.

As long as Fern moves slowly, she easily skirts obstacles; but when a strange dog lets out a howl, her frantic leap topples a pile of books. While an anthropomorphic view of Fern's dilemma might tempt us to rush in and turn on some lights, let's look at the problem through Fern's eyes.

How Much Light Is Enough?

Given our human diurnal habits, we often react to the idea of the night being a hundred million times darker than the day with a feeling of dread. Isaac Asimov plucked this human chord ever so gently in "Nightfall," which many consider the greatest science fiction short story ever written. Yet while Asimov reminds us that our fears of

Nocturnal Behavior: What a Difference a Day Makes

something so natural and normal as darkness could end a civilization, Jules Verne urges us to celebrate the darkness for a very special reason:

> On earth even on the darkest night light never abdicates its rights. It may be subtle and diffuse, but however little light there may be, the eye finally perceives it.

In order to understand what our cats experience, we need to set aside our emotions and see the light in the darkness that so charmed Verne. Try this simple experiment. Go outside some moonless night or observe your dark surroundings as you lie in bed. Of course objects blend into the background more, but if we take time to let our eyes adjust to the reduced light, we discover that even on the darkest night we can see all kinds of things. In my yard I can see where the trees break at the end of the driveway and where we cleared out some brush last summer. The picnic table, wood pile, cold frame, and garden cart look like velvety densities, but because I know what they are, I can easily identify them.

In my own familiar bed I can "see" the beams on the ceiling, the bureau and the wicker hamper. But when I stayed in a hotel while attending a conference in Philadelphia and was awakened by a door slamming in the next room, initially I was so disoriented I couldn't "see" anything in that room at all. Such experiences help us appreciate why a predatory, solitary, and nocturnal animal would tend to limit its activities to a carefully defined and well-known territory.

Unlike humans, Fern doesn't respond quite so emotionally to her new environment. True she does move more cautiously and slowly, perhaps even a bit fearfully because she knows she's vulnerable in her unfamiliar new surroundings. On the other hand, her careful and thorough explorations will generate that familiarity quite rapidly.

Sound Alternatives

Although most behaviorists ascribe a keen sense of hearing to the cat and consider it most critical to its survival in a limited-light environment, less concrete data exist regarding animal hearing versus animal vision. The problem? How do we describe and analyze something that is essentially beyond human comprehension? Because we see more color or detail than cats, we can imagine seeing less by looking at the world through a foggy window; but we can't magnify our hearing so easily. Compared with cats, we're physiologically deaf in much of their normal range. Therefore even if sensitive electronic equipment reveals that cats send and receive sounds in the ultrasonic range, we can't begin to imagine those sounds. Consequently, much of our knowledge of feline hearing comes from our awareness of its effects rather than an understanding of exactly how it functions.

If we view the cat as a total sensory being rather than one composed of discreet packets of sensory anatomy and physiology, we can appreciate how feline hearing compensates for deficits in peripheral visual perception. Our own senses constitute such an integral part of our being, we take them for granted; and this is especially true for hearing. Although we utilize this sense constantly, we rarely consider that hearing really depends on the coordination of two separate components: the hearing apparatus and the sound itself.

Remember the old teaser, "If a tree falls in a deserted forest, does it make a sound?" This puzzler, quite reminiscent of Schrödinger's Cat, reminds us that hearing depends both on the nature of the sound transmitted and the equipment we use to perceive it. In order to understand the complete system, we need to study both components.

Films of wild predatory felines stalking their prey show them gracefully sweeping their heads from side to side. While the eyes look ahead to detect any movement, each ear scans the peripheral areas for sounds of prey or danger. When a suspicious or promising bit of sound information comes along, the cat immediately stops and looks in that direction. This accomplishes several goals. First, it enables the cat to aim its sensory apparatus more specifically. Second, it helps the cat accurately determine its distance from the sound. Third, it lets the cat compare the arrival times of the sound in each ear, thus further pin-

Nocturnal Behavior: What a Difference a Day Makes

pointing the location of the sound. And finally, it allows the cat to hear the sound again: If the initial sound is brief (as most rodent sounds are) the cat needs to hear it again to know more about its nature and origin. By standing stock-still the cat quickly and efficiently collects and analyzes sound data to distinguish identity, location, and the direction and rate of movement if any. The beauty of such a system, of course, lies in its ability to precisely locate and identify moving objects without the benefit of any light.

In addition to this incredibly sensitive anatomical apparatus, cats also possess certain highly specialized brain functions that further enhance their ability to process and use these sound data. One of these involves sound habituation, the ability to increase or decrease a response to certain sounds as a result of experience. A typical example of how this works is demonstrated in the experimental work of Hernandez-Peon, Sherrer, and Jouvet, who discovered they could easily distract a relaxed cat from a previously attractive clicking noise by introducing a live mouse or the odor of fish to the environment. While we might think this reaction indicates that vision or odor takes precedence over hearing as the major sense, the research actually demonstrated that cats were able to dampen the stronger sound stimuli in order to concentrate on weaker, but more promising, sensory cues. You and I do this all the time; even as I concentrate on typing these words—seeing them written in longhand and hearing the sound as I hit the appropriate keys on my typewriter—a faint odor is enough to send me flying to the kitchen to check on the dinner baking in the oven.

The advantages this ability offers limited-light inhabitants are obvious. Many species move about during these periods—some predators, some prey, some inconsequential to the individual's survival. In order to survive, the animal must not only be able to distinguish threatening or promising sounds from others, it must also ignore stronger but inconsequential sounds in favor of lesser but more significant sensory cues. So the cat must be able to ignore the sound of the

THE BODY LANGUAGE AND EMOTION OF CATS

flapping shutter or plastic sheeting on the woodpile and turn her full attention to the faint movement under the porch if necessary.

In addition to hearing and choosing to ignore sounds in their worlds, cats make sounds themselves. We already noted one advantage of sound for a solitary creature: Because sound travels through air, it can wend its way around an intervening object. Even if the visual field remains clear, animals functioning in limited light can hardly rely on visual cues, especially when they lack the ability to detect detail. Equally important, animals employing sound communication may engage in other activities. A feeding animal may continue feeding while it analyzes sounds from its environment; a queen with her eyes fixed on something threatening may simultaneously use her ears to locate her young, then call them to her side.

Like many animals, cats vocalize in response to three behavioral or physiological states. Those who can't escape a confrontation with a perceived predator hiss, growl, snarl, and scream. Competitors or territorial violators also elicit a similar range of vocal signals. Third, a set of distinct vocalizations accompany the mating process. Initially "come hither" calls help the amorous couple locate each other. Then a different set of softer verbal exchanges occur at close range as the felines begin wooing each other. Then there's that awful unearthly shriek many females emit at the end of copulation. At such times the successful suitor may feel the need to vocally repel any other males who happen to be nearby.

While anyone who's encountered angry, frightened, or mating cats may feel that other, more obvious and elaborate body language displays would make all this vocalization unnecessary, bear in mind that such sounds evolved to meet the needs of an animal who conducted most of its activities in limited light. While that sharp-fanged, glazed-eyed spiky-furred daylight image is indelibly imprinted on our human retina, such an image is never seen in such detail by the feline's other foes or potential mates.

Nocturnal Behavior: What a Difference a Day Makes

ORDERLY ODORS AND NOCTURNAL OLFACTION

The sense of smell also comes in handy in limited-light environments. Although we noted previously that smell plays a relatively minor role in the location of prey, it does provide an efficient means of intraspecies communication for territorial species where messages may need to traverse long distances. Like sound, strong scent cues may completely override other environmental stimuli; the putrid discharge from feline anal glands easily takes precedence over any other odor in the vicinity.

Because scent cues pervade the environment and are more difficult to localize than sound or visual signals, they must be durable and intense. Imagine trying to locate someone wearing a particular brand of perfume in a crowded room versus seeing or hearing someone waving frantically or calling your name.

If such scents must be potent and durable to help locate their source, why don't people living in areas with dense cat populations succumb to a cloud of noxious feline scents every breeding season? Because these odors last so long, cats reserve the most pungent ones for the most obviously important or threatening situations. It wouldn't make sense, for example, for a female to produce a scent indicating her willingness to mate that lasted for weeks. Once she's been bred, that message would only attract unwanted suitors to her territory. Consequently, most feline secretions are emitted in small doses and rarely attract the notice of any but their own kind or those with a specific interest in the individual, such as potential predators or prey.

Because odors, like sound, carry on the wind, this form of communication also functions best in a stable population with fixed territories. An animal that inhabits a fixed territory can use weaker, shorter-term scents because it is present in the area to freshen them as needed. This periodic freshening of weaker scent marks not only enables the cat to further establish its territory, but also prevents olfactory fatigue, the scent equivalent of sound habituation. When exposed to a scent, most animals initially respond, but that response gradually wanes until they don't notice the odor at all. Recall entering a room and being over-

The Body Language and Emotion of Cats

whelmed by an occupant's perfume or aftershave. Most of us soon forget all about the odor until someone else enters the room and comments, "Geez, who's wearing Barnyard Number Nine?" So, no matter how potent a scent, the receiver of the message has as much influence over its effect as the sender.

Although little data exist concerning feline scent trails, we do know that members of some nocturnal species use the scent glands in their feet to mark safe and navigable routes through their territories. In some tree-dwelling mammals this scent path is so inviolate that if they should happen to slip or lose their balance while laying down the original trail, they continue to misstep or deviate every time they return to that location. We tend to associate most feline face and head rubbing around the house with "ownership" signals, but in fact such marks often occur in doorways and on prominent pieces of furniture. This raises the possibility that these caresses also function as scent "buoys" to help channel the cat safely through our homes. Remember my brother's cat, who rubbed her face on the coffee table as part of her nightly petting ritual? I can't tell you how many times people stumbled into that same table in poor light, whereas the cat always bypassed it neatly even in total darkness.

The Magnetic Cat

Many owners believe their cats possess a special charm that makes them almost mystical: "Marmalade's such a personable cat, people are drawn to him like a magnet," notes Karen Norton. "You just can't help falling in love with him." While such forces may draw individual cats toward others, all cats are sensitive to minute changes in electromagnetic forces that keep them from running into things. Although we don't understand the exact mechanism involved, it could very well be that the cat's ability to perceive changes in this realm enables it to experience the equivalent of "ultrasonic" touch or pressure, detecting shifts in its environment at the submolecular or even subatomic level. While the idea of sensing the shift of something so minuscule as a

Nocturnal Behavior: What a Difference a Day Makes

molecule may seem ludicrous, all senses, including our own, routinely function at such levels. The photons or units of light that stimulate the retina are even smaller, the size of one electron. Sound travels in waves of air forming particular crests and troughs in the ocean of molecules that makes up our atmosphere.

In the more intimate or immediate environment, objects compress, displace, or repel that ocean of air. If such objects remain stationary, the movement of the air around them becomes as predictable as that of a stream flowing around a submerged rock. If the object moves, it creates a variable pattern that clearly communicates its motion.

In such a way the tactile hairs on the cat's face and front legs apprise it of its immediate surroundings, serving more as an orienting than a locating mechanism. Recall how both the pinnae and the eyes must focus on the prey to locate it with the precision necessary to pounce. Wherever the eyes point, the ears and nose must surely follow. While this creates a sensory concentration that makes the cat a most efficient predator, it also leaves it sensorially deficient and vulnerable to the sides, a weakness that makes those elegant feline whiskers all the more important. Projecting to the sides in a plane perpendicular to the visual plane, this exquisitely sensitive system serves to inform the cat about its immediate environment without distracting it from its target. When limited light makes the exact position of an immobile mop and broom against the kitchen wall difficult to discern visually, Fern's tactile hairs warn, "Move to the left." Similarly tactile hairs in the carpal area tell the cat the relationship of its front paws to its environment. Not only is this critical information to the hunter trying to hang on to its prey, it is equally critical to Fern as she tackles the pitch-dark cellar stairs of her new home for the first time.

TURNING NIGHT INTO DAY

The cat's flexibility may extend into the inviolate realm of night, but many typical problems encountered by owners also testify to the cat's resistance to change. For example, when the Cruikshanks "saved" Fern from a lifetime of fending for herself, she'd already adopted a primarily nighttime existence. Because the queen's territory had consisted of little more than the barn and its immediate environs, and because the rodent population took cover when humans worked there, Fern had inhabited a basically nocturnal world.

"How can that be?" questions Gerri. "We saw Barn Cat nursing and grooming the kittens in broad daylight!" Of course they did, because, as we know, both of these "resting" activities occur when the cat can't hunt. Although Fern did receive more nursing during the daylight hours, she ate her first solid food—a mouse—at night. Furthermore, during this very impressionable period, her first feeble attempts at hunting and territoriality also took place under cover of darkness.

Even though the lessons imprinted at such an early age resist alteration more than those patterns established at a later age, all require some form of reinforcement to maintain. Because of this, owners whose cats go in and out often have a much more difficult time accustoming their pets to a diurnal activity shift later on than those whose pets stay indoors all the time.

The Cruikshanks, like many contemporary cat owners, observed a common human/feline cycle in their old neighborhood. Fern spent the night outdoors, but actually conducted her most active prowling and hunting shortly after her owners let her out and shortly before coming in in the morning. Because Ben and Gerri both work, this schedule functioned well for humans and cat alike. Most of the time Ben and Gerri spent with Fern convinced them that she was quite active and responsive. While they worked, the calico slept contentedly on their bed, so they never worried about her getting into trouble. At night, when she was out, they were home to attend to any problems that might arise.

Then the Cruikshanks' move completely disrupted Fern's schedule. Wisely, her owners kept her indoors, knowing she would be vulnerable until she established her new territory. And in spite of her disrupting their sleep, they were determined not to let her outdoors until they made a careful survey of the neighborhood.

As it turns out, aside from the dog next-door who stays on a leash or in a fenced yard, the new neighborhood contains few other animals or threatening conditions. Their house sits on a dead-end street abutting a woodland similar to the one Fern frequented in her former habitat. Therefore, after introducing Fern to the area gradually and under close supervision, Ben and Gerri quickly integrate her into her new territory without either cat or humans needing to alter their day-night orientations.

However, suppose their survey of the neighborhood revealed several loose dogs—including a known cat hater—a lot of noisy kids on bicycles, heavy traffic, and a landfill just over the hill. In that case, the Cruikshanks might very well decide to convert Fern into a house cat and reset her biological clock.

Whether or not owners succeed in such endeavors depends on the strength of their belief that this course offers the best benefits for both them and their cat. Because Ben and Gerri feel Fern will be safer indoors, they find it easier to ignore her and not become angry when she disturbs them during the night. In those first weeks of often-interrupted sleep, they force themselves to respond cheerfully and actively to Fern in the mornings and evenings when they're home, thereby teasing her biological clock to align itself more compatibly with their own. Although it takes several months of consistent effort, eventually Fern and the Cruikshanks achieve a sufficiently similar rhythm to maintain a solid bond.

What about nocturnal fighting and howling? Because most of this commotion attends mating, trying to eliminate this behaviorally proves most unrewarding. However these displays do respond to surgical treatment. Simply spaying the females or castrating the males will safely and permanently eliminate this most troublesome nocturnal display in most animals. To be sure, such surgery doesn't cause cats instantly to abandon their nighttime orientation, but it does eliminate one of the strongest instinctive drives that propels domestic cats toward greater activity in limited light.

THE EXCEPTION TO THE RULE

Before considering the second most common problem related to the lack of synchrony between human and feline biological and behavioral cycles, we need to consider that long-standing feline exception to the rule, the Siamese mutation. What difference does color make? Wasn't Ben Franklin right when he noted, "At night all cats are gray"? In terms of some specific variations in anatomy and physiology and the effects these have on feline relationships with humans, the Siamese mutations, their relatives, and descendants do *not* respond to their environment the way other cats do. If we forget that such variations exist, we may try to force our cats into behavioral patterns quite alien to their normal bent or, worse, judge their perfectly normal behavior as wrong or inferior.

Although we'll be discussing the familiar blue-eyed mutation, bear in mind that during the hundreds of years since the complex mutation first emerged, it has assumed many forms. (The solid sable, golden-eyed Burmese seems remote from its blue-eyed black-and-white ancestor, yet its strong people-orientation readily parallels that of the Siamese.) More than one owner who finds his or her cat's problem behavior incomprehensible when compared with the mackerel tabby in Uncle Harry's barn may find it perfectly understandable when compared with Aunt Helen's Himalayan. Let's review some of the basic characteristics of the Siamese mutation, then see what specific implications they may hold for nocturnal behavioral displays.

We noted that the blue eyes and white coat with its associated point coloration indicate some degree of albinism in the early Siamese mutation. We also noted that the coloration of these animals makes them simultaneously more appealing to people and more vulnerable than their dull-brown compatriots.

Although we can amass little hard data on the subject, we know that

Nocturnal Behavior: What a Difference a Day Makes

blue-eyed animals lack a reflective tapetum and, unless they possess some completely different and unknown light-intensifying mechanism, they most likely possess limited night vision when compared with other cats. Furthermore, we also know that deafness often afflicts blue-eyed, white-coated cats. Finally, a quick survey of the blue-eyed cats of Siamese origin—Siamese, Himalayan, Birman, Balinese, Burmese, Javanese, Ragdoll, Colorpoint Shorthair—reveals a high degree of vocalization and human sociability.

While most of us tend to associate vocalization with these cats' lovable natures and desires to communicate with people, we can't overlook the possibility that it initially resulted from an attempt to compensate for the diluted sensory responses that accompanied the diluted coloration. Recall how cats suddenly deprived of vision or hearing as a result of trauma or disease often begin vocalizing. Although we can't be sure whether such vocalization enables the cat to orient itself via some mechanism beyond our comprehension, we do know that such cries inevitably bring humans running. And if they come bearing food, so much the better.

Moreover, when humans partially lose their hearing, they most often lose their ability to hear the higher frequency sounds first because these activate the most delicate and easily damaged auditory tissues. Suppose that whatever factor triggered the Siamese mutation also affected its ability to perceive sounds in the upper ranges, and perhaps volume to some extent as well. Would such animals then communicate both at a lower pitch and more loudly than their unaffected peers? If this were the case, we can see that these cats would lose out on two fronts relative to their ability to survive in a limited-light environment. First, they wouldn't be able to hear in their prey's normal range of communication. Second, they themselves would be vocalizing in a range that could draw a broader selection of predators to them.

When I recall my own Siamese and Himalayans and those I have seen in practice, it seems that only the most robust, confident males expressed much desire for night life, and even these displayed much more interest in other cats than in hunting. Most of them preferred to sun themselves in broad daylight and preferably within yowling distance of their owners. Those few that indulged in any hunting at all seemed to be diurnal, preying on birds with limited success.

This observation poses another dilemma for us humans. If we view certain physiological changes from a strictly behavioral point of view, we can't help but marvel at the impact humans have made on the development of the modern cat. Regardless of the physiological function of vocalization or the visual or auditory capacities of those first feline mutants humans brought out of a limited-light environment into the full light of day, those characteristics confer many appealing qualities on the species. Although on the one hand we may disdain this as dastardly human interference, on the other it serves as a fine testimony to the cat's genetic and physiological flexibility.

Furthermore, the evolution of increased lower-frequency vocalization associated with decreased sensory abilities makes one wonder about language as a "superior" human achievement. Is it possible that, like those early blue-eyed semi-albino feline mutations, we humans developed our vocal skills not because our brains were so exceptional but because our senses were so limited? Because we couldn't see, hear, and smell each other to our satisfaction, we began yelling at each other instead?

Although a new resident who had never seen Marmalade would have little difficulty attributing Siamese breeding to him based solely on those pitiful yowls he emits as he huddles outdoors, such a possibility never enters Bruce Norton's mind. All he sees is a big brawny orange tiger who ought to be in his glory at night instead of crying like a big baby.

TURNING NIGHT INTO DAY

For all the havoc caterwauling toms and amorous females can wreak on human sleep, it's a lot harder to turn a daylight cat into a nocturnal one than the other way around. The daylight cat usually results from close human interaction, which makes its people more important to it than its territory. Needless to say, cats who develop strong people associations from an early age rarely develop strong nocturnal or predatory behaviors, because these usually are neither necessary nor rewarded. Consequently these cats will contentedly pattern their life-styles to complement their owners'.

People-loving daylight cats are usually born into loving, often sheltered human environments, often to queens viewed anthropomorphically by their owners. For example, Marmalade and his two littermates enjoyed as much contact with people as they did with their mother. Although the queen was a good mother and was devoted to her kittens, she also worshiped the woman who fed and groomed her—and then her kittens—daily.

Consequently, the same personality traits that enable Marmalade to gravitate so readily toward the Nortons also guarantee his unhappiness without them. While the young Marmalade doesn't mind spending the nights outdoors sleeping on the Nortons' porch, the combined effect of being neutered and the introduction of several new, fully feline toms in the neighborhood makes such an "unhuman" existence not only distasteful, but also threatening to him.

Once again we encounter a situation requiring that the owners choose whether they want to spend the time necessary to override some very deep and powerful behavioral patterns. In this case Marmalade's neutered status makes the cat-dominated world of the night even less tantalizing than it was before; and, seeing more reason to say indoors, he spends more time and effort seeking out hiding places that will enable him to do so.

To further complicate this problem, Karen and the children encourage Marmalade's close interactions with them. Even though Karen adheres to the careful search procedure set down by her husband, she secretly cheers when Marmalade escapes detection. However, whenever this occurs, she lies awake praying he won't jump on the bed until Bruce falls sound asleep.

THE BODY LANGUAGE AND EMOTION OF CATS

At this point we see the owners and cat perched at the apex of a vicious cycle that can do nothing but undermine their relationship. If Bruce feels he "wins" every time he succeeds in putting Marmalade out and the cat sees this as a "loss," human and cat can never achieve harmony. Given a lack of harmony and the inconsistent response of other family members, it makes much more sense for Bruce to accept Marmalade for what he is and learn to enjoy this special rhythm of his cat's biological clock.

What if Bruce can't accept this behavior? What if he believes that cats should stay out all night or that they spread germs or cause allergies if they're allowed on the bed at night? If we consider our own beliefs or fears unchangeable and in conflict with our cat's behavior, we do ourselves and the cat a favor if we terminate the relationship. Neither Marmalade nor the Nortons derive anything of value from a relationship based on a daily battle of wills. Such tension and stress can only lead to more serious behavioral and possibly even medical problems.

Although the Nortons might try several available solutions to their problem, they stumble serendipitously onto a unique one. One of their children brings home a stray cat that has been hanging around the school yard for several weeks. Because this animal knows how to fend for himself, he exemplifies what Bruce believes a "normal" cat should be; he *wants* to go out nights and sleep all day. Now the Nortons follow a completely different ritual following the eleven o'clock news. As the weatherman sums up his forecast, the new cat yawns, stretches, and gravitates toward Bruce for some rubbing and purring. Meanwhile Marmalade gives his ball a few more bats and looks from Karen to the stairs. As soon as the news ends and the Nortons rise, the stray streaks to the door to be let out while Marmalade bolts for the bed to secure his favorite spot between Bruce and Karen. Because both cats are affectionate and well-behaved in entirely different ways, the Nortons come to appreciate and accept a much broader range of "normal" feline behavior. A novel solution perhaps,

Nocturnal Behavior: What a Difference a Day Makes

but one totally befitting the unique relationship between human and cat.

During our discussion of nocturnal behavior, we mentioned the role territoriality plays in the cat's ability to function successfully in a world of limited light, but one cat fancier, Hal Goldstein, can't overlook the territorial behavior of his pampered feline, Toulouse, either. After an exemplary kittenhood, Toulouse suddenly begins urinating in Hal's closet and sharpening his claws on the drapes. In the next chapter, we'll try to repair the damage.

6

TERRITORIALITY: TO HAVE AND TO HOLD

$\mathcal{T}$OULOUSE the black Persian enters Hal Goldstein's life in a manner that greatly affects their relationship from the very beginning. Hal, an interior designer, stops to visit a former client to see how she likes her new decor. The homeowner greets him clutching a furry black ball in her outstretched hands, tears streaming down her face.

"Look at him, Hal," she sobs as she thrusts the kitten toward him. "He was supposed to be the foundation of my new line." The kitten looks perfectly normal and healthy to Hal as it playfully bats at his finger. "Why in the world is she so upset?" he wonders.

Suddenly the woman stamps her foot angrily. "He's much too small, his eyes are the wrong color, and he's just, well, too *loose*. Here, take him. Get him out of my sight!" She hangs the kitten on Hal's loosely woven sweater like a furry campaign button and slams the door. Dazed, Hal stumbles to his car, the kitten clinging to him for dear life.

Given such a dramatic introduction, the relationship between Hal and the Persian takes on a special quality from the outset. Because Hal's professional eye gears him to evaluate objects in terms of proportion, he examines the kitten carefully to see if he can discern what made his client so hysterical.

"Well, you look perfectly normal to me," he tells the kitten, who seals the friendship by placing one cool black foot pad on Hal's nose.

Territoriality: To Have and To Hold

"So you're too 'loose,' eh? Then I guess that makes you Toulouse."

The relationship flourishes even after Hal begins to detect a peculiar odor in his home. Initially he chalks it up to a very damp spring and his two-hundred-year-old house. Continuing his usual schedule, he hosts a number of intimate dinner parties. However, he catches several guests wrinkling their noses as they enter his home, and as Toulouse approaches his first birthday, the odor becomes unbearable. Because Hal plans a gala party to celebrate the event, he decides to have professional cleaners scour his home from top to bottom. "I don't care what it takes," he instructs the work crew as he prepares to leave for work. "Just find that stink and get rid of it."

The cleaners accept the mission, setting about their work with a vengeance and scaring poor Toulouse half to death with their vacuum cleaners, buckets, brushes and sprays. "Oh, oh, he's not going to like this," groans the worker who discovers the urine-soaked corner of the closet in the spare bedroom. He gingerly picks up several pairs of shoes and some sports equipment, discolored and reeking of urine. The crew foreman leaves a note for Hal detailing the condition of the closet, but the first thing Hal notices when he returns are the shredded custom-made draperies on the front window.

Until the Clarks adopted her, Reject's life read like something out of a Dickens novel. Born at a private, poorly funded animal shelter, she experienced little outside her cage but a few pats twice a day from her harried caretaker. When she and her littermates were only a few weeks old, their semiwild mother escaped from the cage and disappeared. Then the caretaker became seriously ill and had to close the shelter, shipping Reject and her three littermates to an already overcrowded public humane society, where they huddled together at the back of yet another cage. Two of the kittens died, another finally found a home, leaving only poor Reject trembling and alone in the cage.

The Clarks actually had no intention of adopting a pet, but Reject's pitiful appearance wins their hearts. Barely twelve weeks old when she enters the Clarks' home, several months of loving care must pass before she will accept petting from the Clarks and play even minimally with them. Although her attitude toward them improves dramatically, the instant a stranger appears, she flees; if escape proves impossible, she hisses and lashes out with her claws. Compared with her behavior when she first entered the household, her present state strikes the Clarks as marvelous. Their attitude alters dramatically when they receive notice that they are about to become parents: the long-awaited approval for their adoption of a three-year-old Asian orphan finally arrives.

Our discussion of nocturnal behavior revealed the reasons behind some remarkably varied and flexible feline displays. However, unlike circadian rhythm, territoriality recognizes few natural variables, such as lengthening or shortening days, which can breathe flexibility into feline responses. The large birch that marks the boundary of a calico's domain remains a functional sentinel whether clothed in summer's leaves or stark naked in the December cold. Given such fixed markers, we might think that a study of territoriality would require simply that we learn the rules of the game and how to apply them: "The average house cat requires x square feet of living space and y accessible levels within that space." Unfortunately, that's not the case.

Although we can discover such specific requirements among the data compiled by scientists observing animals under strict laboratory conditions, few cats live in such orderly unchanging environments in the real world. Many veterinarians and behaviorists routinely caution clients about the hazards involved in introducing a second or third cat into a household because of the feline's solitary nature. Time and time again we hear of owners who accomplish this feat with no preparation and no problems whatsoever. Others follow our advice to the letter, only to suffer every conceivable form of negative feline behavior.

What causes two members of the same species and family to engage in all-out feline warfare while two others play and sleep together contentedly? To answer that question, we must abandon any rigid beliefs we've developed about our highly flexible feline friends.

Territoriality: To Have and To Hold

A MATTER OF DEFINITIONS

As we seek to understand territoriality, we run smack up against our own, usually very strong, personal beliefs about social versus solitary behavior. How much territory or space does an individual really require? Consider the views of some illustrious sages:

> "With few desires live alone and do no evil; like an elephant in the forest roaming at will."
>
> Anaxagoras (ca. 500–400 B.C.)

> "It is not good that the man should be alone."
>
> Genesis 2:18

> "We are for the most part more lonely when we go abroad among men than when we stay in our chambers. A man thinking and working is always alone, let him be where he will.
>
> Henry David Thoreau (1854)

> "You don't live in a world all alone. Your brothers are here too."
>
> Albert Schweitzer (1952)

Given such widely accepted but divergent human standards, we can readily see that even though behaviorists may define humans as a social species, in truth many disagree with that notion—at least on the surface. Although Anaxagoras pleaded for solitude, he cited kinship with the elephant, which, among mammals, maintains some of the strongest social ties and roams singularly at will less than many people.

Even though we often hold strong beliefs regarding the sanctity of private or solitary versus social behavior, in reality we may be considering two different aspects of the same behavior. How can we define solitude without comparing it to companionship? And can we discuss gregariousness without taking its opposite into account? Behold another paradox: We can't be alone except in terms of our relationship with others, and we can't be with others except in relationship to our aloneness. Unless we can accept this paradox, we will foist an unworkable dichotomy onto our pets. All individuals of all species may periodically feel the need to be alone, just as they may periodically feel the need to be with others of their own or another kind.

Once again, we must come to grips with the fact that our personal beliefs play a critical role in our interpretation of animal behavior. While Hal delights in Toulouse's rushing to greet him and interact the instant he enters the house, one visitor to his home notices the Persian's arrogant aloofness while another complains about the cat's defensive hiss when she tries to pet him. Although all three cases involve one cat greeting one person, the relationship between the cat and each of these people differs markedly. Not only does Toulouse view each human differently, each human holds different perceptions about the cat's behavior and what it means. All of these views include more than behavioral or psychological relationships. They also depend on the amount and quality of the physical space that separates those attempting to relate to one another.

Consequently we can't talk about territorial behavior without discussing social and asocial orientations; but before we can do that, we need to clarify some terms. In some scientific circles the term *solitary behavior* has replaced *asocial behavior because too many people equated* asocial (literally, "without a social structure") with antisocial behavior, which indicates antagonism or hostility. The highly limited "If you can't be for us, you must be against us" view evoked by this erroneous association of asocial behavior led some researchers to drop the term in favor of the less judgmental and emotionally charged *solitary*. Because both *asocial* and *solitary* do appear throughout the literature, we'll use them interchangeably. However, remind yourself if necessary that *asocial* doesn't mean "antisocial."

What do social or asocial orientations have to do with our discussion of territoriality? Quite simply, both describe the *psychological* space or territory an individual needs to feel secure. All animals possess physical space needs that are critical to their survival; these ensure adequate food, limit predation by other species, and provide mating opportuni-

Territoriality: To Have and To Hold

ties. The amount of land, air, or sea necessary to fulfill these needs constitutes a given animal's physical territory. A wild feline dwelling in a lush forest or meadowland supporting a large prey population needs less physical space to fulfill its needs than one living in a harsh arid environment. A barn cat, which depends on mice for food, requires more space than the house cat whose food dish is always full.

In addition, each animal also requires personal space, which may be anything from relatively minimal (such as that of subordinate members of a large herd of grazing animals) to extensive (such as that of some male wild cats who may only tolerate the presence of their own kind when mating). If an individual requires minimal personal space, we call it a *social* creature. If it requires a great deal of personal space, we call it *asocial* or *solitary*. In such a way the social or solitary orientation determines the psychological territorial needs.

While these theoretical concepts appear quite logical, we need only consider our own orientations and those of our cats to realize that the exceptions are as numerous as the adherents to the rule. As a human, I'm a member of a social species, yet I do my grocery shopping at 6:30 A.M. to avoid my fellow humans; my personal and physical space needs in a supermarket are much greater than when I entertain loved ones in my tiny home. Maggie belongs to the asocial feline species, yet at times she's so desirous of companionship that she practically sits on top of humans or dogs. So although the concepts of territoriality appear both logical and predictable, the capricious natures of both human and feline allow for infinite variations within each orientation.

THE PHYSICAL SPACE

In order to understand how an individual cat responds to a specific set of physical conditions, we need to understand the basic functions of the physical territory. Obviously territoriality must confer some survival benefits, otherwise cats would never have incorporated it into the fabric of their existence. Just as periods of wakefulness and sleep efficiently divide the day, so territoriality effectively divides the physical space.

THE BODY LANGUAGE AND EMOTION OF CATS

In our discussion of nocturnal behavior we noted the crucial role fixed territory plays in the feline predator's ability to move quickly and quietly when locating and capturing prey. In such a way established territories help guarantee a reliable food source. The exquisite interplay between the social rodent prey and asocial feline predator further enhances this benefit because the cat's asocial nature guarantees that fewer predators will stalk a given territory, competing for the food supply; the social nature of rodents simultaneously guarantees there will most likely always be more prey than predators in a given area.

Defined physical spaces also serve to control population. It takes more effort for a receptive female in one territory to communicate her status to a male in another territory than it does for a female who lives with a pack or herd. On the other hand, individuals displaying intense territoriality who do so at the expense of mating will not contribute to the gene pool of the species. In such a way creatures balance the need to stay at home (asocial) against the need to roam and mate (social). Although Toulouse rarely shows any desire to leave the house, when he does slip out, Hal's immediate response sounds familiar to many owners: "There must be a female in heat somewhere in the neighborhood."

Fixed territories also provide more efficiently for defense. In this capacity, the physical space not only awards the individual maximum knowledge of its territory but also minimizes hostile contact with members of its own species. In such a way, animals avoid wasting valuable energy on intraspecies fighting. Similarly these individuals can move more readily when defending themselves against predators. Even though the predator may be bigger and stronger, intimate knowledge of a specific physical space reveals all sorts of hiding places and escape routes not readily recognized by a less knowledgeable intruder. The Clarks experience little difficulty locating Reject in their own home once they discover her favorite hiding places. However, when an elderly neighbor asks them to catch her cat so she can medicate it, the task seems hopeless. Not only are the Clarks unfamiliar with the layout and furnishings of their neighbor's home, they have no idea of the kinds of hiding places her cat prefers. Does the cat prefer to crouch under low furniture like Reject, or is she one of those cats who likes to hide on top of cabinets or in drawers or closets?

Another benefit of defined physical territories involves disease control. Animals dwelling in specific spaces exercise greater care when depositing their wastes. Solitary wild cats bury their wastes, thus reducing the chance of infecting themselves or their own kind with disease or parasites as well as eliminating identifying scents. Furthermore, most wild animals deposit their wastes away from feeding and sleeping areas. This can be such a strong instinct that cats whose food and water bowls sit too close to their litterboxes may refuse to use the latter. Similarly, an owner can discourage cats who begin urinating and defecating in inappropriate places by placing their pets' food dishes on the spot. Nevertheless, because the benefits conferred by burying wastes no longer play a critical role in the health and welfare of many domestic cats whose owners assume these responsibilities, fewer cats honor the tradition.

The benefits of territoriality for specific individuals fall mainly in the psychological realm. An animal claiming a specific territory functions more decisively and confidently than one functioning on changing or unfamiliar ground. Secondly, animals defending a well-known territory also receive a psychological boost conferred by possession. Whether possession constitutes nine-tenths of the animal law or not, an animal challenged on its home ground will mount a far more vigorous defense, willingly taking on bigger and stronger adversaries with a wider range of tactics for a longer time, than one challenged in a foreign territory.

This awareness carries its own paradox, familiar to many veterinarians and groomers. On the one hand, the cat on its home turf is more confident and therefore, in theory at least, should be more tolerant of examination, handling, and treatment in this environment. Hence some owners and professionals maintain that animals treated at home do better and that house-call services best meet the animals' and owners' needs. On the other hand, defensive territorial animals are much more likely to display such behaviors in their own homes, thereby making any examinations, handling, or treatment much more difficult under these circumstances. Because of this, we must recognize our own cat's orientation in order to select the most appropriate form of care.

THE PERSONAL SPACE

Differentiating personal from physical space takes a lot of subjectivity and intuition. When Heini Hediger observed birds sitting on a Zurich fence half a century ago, their space requirements seemed fairly obvious: The flamingos needed two feet of space between them to settle peacefully, the black-headed gulls required a foot, and the swallows six inches. Although it's tempting to classify such space requirements as highly territorial in nature, they obviously don't relate to food supply or reproduction. A flamingo certainly needs a larger area to fulfill its nutritional needs; and two flamingos could hardly mate on twenty-four inches of fence rail. Compare these birds' behavior with that of redwing blackbirds arranged along fencerows in farming country. Here the much greater distance between birds easily meets the criteria for physical territory. For both flamingos and redwing blackbirds food supply dictates the different forms the personal and physical space assumes. Because flamingos, gulls, and swallows share a common hunting space with other members of their own species, their maximum displays serve to establish and protect their personal space; but because the redwing blackbird hunts in an exclusive territory, its personal and physical territory are one.

Cats also align themselves in similarly contradictory manners. People in multiple-cat households often recognize specific space orientations among their pets: "Muggy lets Pumpkin sleep next to her, but never Abercrombie. If Stardust even comes into the room, she attacks him." However, even though such specific requirements may govern sleeping or play space, all cats may contentedly eat from the same bowl. In other settings, cats may function literally as guard dogs; if any intruder of any species comes into the yard, these cats will attack.

Although birds draw fairly clear lines between physical and personal spaces, other species do not. A pile of tortoises sunning themselves or a pile of exhausted wild pups or kittens seem to require no personal space at all. A year later, however, the tortoises are still piling up but the pups live within an established space with clear personal space requirements, and the cats now reside in individual physical territories that exclude most interactions with their own kind. Reject clung to her littermates as though life itself depended on such behavior; as an adult, she views the presence of another cat as life-threatening.

Personal space tosses a behavioral wild card into the game of life. Even in a relatively orderly natural environment it doesn't play out its hand the same way every day. For example, a wild cat that doesn't normally distinguish between its physical and personal space with respect to other members of its species will only apply those limits to members of the same sex during breeding seasons. Males may drive out other males even more aggressively the instant they cross the physical boundaries, but welcome receptive females. When queens are raising young, their personal space initially includes their kittens, but slowly expands to exclude them. During the period when the queen shares the most intimate contact with her young, she's most intolerant of intruders; but as she begins excluding her kittens, stops nursing, and comes into heat again, her tolerance of others increases. This flexible expansion and contraction of personal space requirements within the physical territory permits both mating and raising of young without permanently increasing the number of adults vying for the same food supply.

PACK BEHAVIOR IN CATS

Aside from lions, the notion of feline pack or social behavior would appear ridiculous. However, a general rule of biology does allow for this apparent behavioral paradox: The more limited the space, the more the rules of dominance and submission (that is, pack behavior) come into play. Thus, even though adult cats lack the wide repertoire of body language cues that signal relative position in highly pack-oriented species, social rules do govern the close interactions of kittens among their littermates and with their mothers. Surprisingly, the members of feline litters do normally distinguish themselves as leaders and followers, as dominant or submissive with respect to others. When the time comes for these individuals to establish their own territories, the lessons learned in this social environment come in handy. When the maturing animals seek an exclusive territory, they must first locate such a place, and doing so may involve numerous encounters with healthy individuals protecting their own territories as well as unhealthy or aged ones a rival might successfully drive off. Either way, those kittens who learned to act the most dominant in the "pack" will most likely possess the self-confidence and independence to successfully establish and maintain a territory.

Ironically, it appears that the characteristics that best enable cats to challenge each other function least when they struggle for survival in the physical territory. Imagine two cats squaring off for combat, their coats puffed up to make them appear larger, emitting threatening hisses and growls, and displaying stiff-legged postures. Now imagine that same animal, sleek and low to the ground, soundlessly stalking its prey. Needless to say, the cat who can master both displays exhibits a broad range of physiological and behavioral expressions. And although the intraspecies signals display neither the subtlety, the variety, nor the complexity of those favored by more social animals, they nicely serve solitary creatures who interact primarily in limited light.

Finding Our Places in the Feline Territory

The balanced interplay between physical and personal space and normal social and asocial instincts fluctuates wildly when domestication enters the picture. Because cats tend to relate to some people the way kittens relate to their queen, the liaison between human and feline can cause arrested or distorted feline behavioral development. In some sit-

Territoriality: To Have and To Hold

uations the cat may continue to behave like a kitten; in others, the adult instincts win out. Many times, however, we wind up with a little bit of each, a hodgepodge of social and asocial behaviors and variable territorial responses.

Let's reconsider some typical feline greeting displays. We noted how house cat Toulouse rushes to greet Hal Goldstein the instant the Persian catches sight of his owner. A lesser but still enthusiastic version of this display also greets frequent visitors to the Goldstein home. When these people enter the house, Toulouse experiences no violation of physical or personal space; they would have to cause him some sort of physical discomfort or otherwise threaten him to elicit any sort of defensive response. When strangers arrive, Toulouse preserves his personal space but allows these others to move freely about the rest of his physical territory (that is, Hal's house). Although neither defensive nor antagonistic, he does maintain his distance. If unfamiliar individuals attempt to force themselves on Toulouse before he feels ready to accept them, he moves away; if he can't flee and they continue violating his personal space, he responds defensively. Finally, if something about the visitor strikes him as sufficiently threatening—such as Hal's neighbor arriving with her brawny macho cat in her arms—Toulouse may decide he can't even tolerate them in his physical territory, that is, anywhere in the house. In this situation, he mounts a full defensive display.

If Toulouse responded to people the way wild cats respond to feline intruders, all of these entrances would elicit defensive displays the instant his physical boundaries were violated. In such cases, physical and personal space function as one. However, Toulouse's responses to people run the full gamut from the strongly social to the strongly asocial. Balancing Toulouse's response to the threatening intruder, where his personal space expands to include the entire house, is his highly social and intimate relationship with Hal, where no sense of personal or physical space separates them.

Between these two extremes Toulouse responds with a blend of social and asocial behaviors, maintaining an insulating sphere of personal space the diameter of which shrinks or expands according to how much he trusts the visitor. He allows trusted ones to move freely, albeit usually under close scrutiny, and he makes defensive displays only when they violate his personal space or get too close to sacred physical spaces or objects such as his eating and sleeping areas or his toys.

Although we can never know for sure what passes through Toulouse's mind, his behavior does appear to parallel that of his undomesticated relatives to some extent. In addition to responding to strangers the way he would to a cat he considers an intruder, Toulouse also responds to Hal as a kitten responds to its queen. More familiar humans are treated more as littermates with dominant and subordinate positions, depending on the degree to which the cat trusts them. Less familiar but nonthreatening people evoke a degree of tolerance similar to what a wild creature affords members of other species in its territory.

In the Clark household, Reject's territorial responses span an entirely different and much narrower spectrum. Whereas Toulouse's sense of personal and physical space expands or contracts over a relatively broad range to accommodate a wide variety of changes in Hal's home, Reject's definitions of adequate territory includes only the Clarks. And even though she accepts the Clarks, she doesn't always accept them to the same degree. Some days she tolerates their presence while she eats or allows them to brush her; other days she hisses when they make the identical displays.

TROUBLE ON THE HOMEFRONT

Earlier we talked about the anthropomorphic or "love me, love my cat" orientation. Many times a variation on this theme causes some serious problems. If an owner enjoys the company of another cat lover and his or her cat, both often expect all cats and people involved in the liaison to get along. This false expectation leads Hal to invite his neighbor, Denise Kuzach, and her cat Interface over for dinner. Unlike house cat Toulouse, Interface roams freely outdoors, coming and going as he pleases through his cat door.

When the two owners hatch this plan, they assume that the two cats will make perfect playmates because both are Persians and males. Interface is a shaded male and about four months older than Toulouse, but aside from these two differences the two cats are remarkably similar in temperament. Of course, this sort of idealistic thinking gets many parents and matchmakers into trouble, and cat owners often fare no better—and often much worse. "Hate at first sight" best describes the encounter between Toulouse and Interface. The shaded Persian leaps fully prepared to do battle, unmindful of the deep scratches he leaves on his owner's arms in the process. Terrified, Toulouse panics and bolts, upsetting a delicately carved pedestal table and Hal's collection of porcelain miniatures in the process. It takes Hal over an hour to capture and evict Interface, calm Denise, and attend to her wounds. Several more hours elapse before Hal finally locates Toulouse huddled in the spare-bedroom closet. Soothing and stroking his pet, Hal promises never to bring the mean bully into the house again.

Because Interface roams freely and Hal owns no other pets, the cat easily claims the Goldstein property as part of his territory. Consequently his daily rounds have always included Hal's yard. However, once Interface realizes that another cat lives inside the house, an almost-mature male no less, he redoubles his patrols and vigilance; and because Toulouse's house sits right in the middle of his territory, Interface marks his territory liberally to make his point quite clear.

Meanwhile Toulouse realizes that what he once believed to be an infinite, safe space he shared with Hal is actually an island surrounded by enemy territory. Particularly when Hal leaves him alone, Toulouse experiences fear and frustration. His kitten instincts, reinforced by his close relationship with Hal and other people, tell him to run and hide; let someone else take care of this intruder. Conversely, his adult instincts tell him he should both mark and protect his territory.

LOST IN SPACE

Across town, Reject's sense of place creates a different but equally dramatic dilemma for both cat and owners. Countering the psychological effects of Reject's early experiences initially proves a challenging and rewarding experience for the childless Clarks. The fact that they must sit perfectly still and wait for the kitten to approach them or that they must refrain from making any sudden movements or loud noises in her presence seems a small sacrifice compared with the happiness they feel when they see her playfully batting her ball. Nor do they resent their choice to at least temporarily halt their previously intense at-home entertainment schedule after Reject badly bites one of their guests, then succumbs to a violent stress-induced digestive upset following a party. This, too, appears a minor inconvenience compared with the joy they feel about her achievements. "Is it possible she's grown and changed so much in the short time she's been with us?" they marvel. "She certainly must be an extraordinary cat to bounce back so quickly," they tell each other. Their joy over Reject's remarkable advances in their custom-made environment pales, however, when they learn that a three-year-old will join their family in a few short weeks.

In this situation Reject's territorial instincts have been warped by her early experiences. Because she spent her critical learning period confined to cages under less than ideal conditions, her awareness of both personal and physical space is limited to an extremely small area. Deprived of her mother and littermates at an early age, Reject lost not only her companions and the sources of social interaction but also her reference points and teachers. During the period when she should have been learning the most about the world, she learned the least. During the time when a kitten normally collects experiences and builds confidence, Reject accumulated one fear on top of another. As a result, no balance or flexibility exists between Reject's personal and physical space and her social and asocial natures. What little she knows resists change; what she doesn't know she fears.

Territoriality: To Have and To Hold

SETTING THE LIMITS

If our cats expand and contract their personal and physical spaces to accommodate or exclude people, how can we avoid inadvertently violating those limits, frightening the animal, and perhaps setting ourselves up for a nasty encounter? The truth is, we can never be sure what limits cats will impose upon a particular environment or individual, but careful observation will usually reveal their spatial orientation toward their owners.

"But Toulouse lets me do anything I want," protests Hal. Exactly. Because Toulouse completely trusts Hal and no boundaries exist between them, the cat need not set any limits. However, the Clarks know that if they approach Reject from behind or get closer than two feet to her without moving very slowly and carefully, their cat will mount a full fearful display. If she can't run, she becomes extremely defensive—ears flattened, pupils dilated, hair erect, crouched low to the ground, and ready to lash out with her claws at the least provocation. And this represents an improvement! When they first got her, regardless of what tactics they used, Reject refused to allow them within the perimeter of her two-foot-diameter invisible "cage" world.

Prior to Interface's visit, Toulouse marked his physical territory with scents beyond human perception. In addition to face rubbing, he scratched and stretched at his scratching post, pressing the pads of his forepaws into the fabric and depositing scent. As a kitten, he attempted this same scratch-and-stretch approach to establish visual and scent marks on several doorframes and prominent pieces of furniture, but Hal covered these areas with double-sided sticky tape and squirted Toulouse with a water pistol when he caught him in the act. This effectively halted the inappropriate scratching; although Toulouse continues to stretch in these same areas and most likely deposits scent marks, he does not scratch.

When Interface violates his territory, Toulouse quickly upgrades and intensifies his marking system. The day the strange cleaning crew arrives tips the scale even more, until instinct overrides learned behavior. As Toulouse crouches in the living room surrounded by the whine of vacuum cleaners and floor polishers, he spies Interface arrogantly striding across Hal's front porch. It's more than Toulouse can bear; he rushes to the window and climbs the drapes in a blind rage. The other, larger cat stops dead in his tracks, then suddenly lunges at the window, taking Toulouse completely by surprise. Toulouse drops to the floor like a rock, pulling down part of the drapes with him. When he's convinced the other cat has moved on, he claws the remaining drape to mark the area, then bolts for the safety of the spare bedroom.

Meanwhile outdoors, Interface backs up against strategically placed fence posts, shrubbery, and doors on Hal's property, spraying them all with pungent urine. He expands the areas where he usually leaves visual territorial markers, clawing the back of the brand-new redwood chaise lounge on Hal's patio. Then he wanders into the garage, spraying Hal's tools and clawing his golf bag for good measure.

Feeling rather cocky, Interface decides to see what's going on in the next yard. He knows a battle-scarred old tom lives there, but perhaps today will be his lucky day. After all, didn't he show that young upstart who's boss of this territory? The fight is violent but short, and Interface quickly retreats to Hal's yard to lick the tiny puncture wounds in shoulder and forepaw left by his adversary's fangs.

Slipping quietly through his cat door, Interface next commences to augment and expand his territorial markers within the house. Spying his beloved mistress's handbag on the floor next to the bed, he wonders, "How could I have missed something so important?" He also sprays the refrigerator as well as a tantalizing human-scented pile of clothes. His rounds end abruptly when, in his zeal, he sprays his owner's back as she lies reading the paper on the living room floor.

Symptoms and Signals

As Reject, Toulouse, and Interface establish, maintain, or fortify their territories, we see a number of familiar displays. Urine and secretions from glands in the head area and foot pads provide potent scent marks;

clawing leaves long-lasting visual signals of territorial possession. When someone violates the cat's personal space, hissing and growling deliver unmistakable auditory messages, enhanced by the cat's distinguishable offensive or defensive appearance. If pushed beyond this point, the cat may evacuate bowels, bladder, and anal glands, adding potent scent signals few could ignore.

Although reading such body language depends on owner experience, inappropriate urination may signal medical as well as behavioral problems. Recall our two Persians: Interface sprays doors, piles of clothing, handbags, whereas Toulouse urinates only in the spare-bedroom closet. How tempting it would be to attribute all the displays to variations on the same territorial theme, but such could dangerously mislead us. When Toulouse first sought out the safety of the spare-bedroom closet, his fear caused the young male to urinate in the corner. Because urine serves as a "sanitary" marker as well as a territorial one, Toulouse returns to the closet, using the area as a second litterbox.

However, the stress of his confrontation with Interface also creates another form of violation—this one a physical violation; Toulouse falls prey to a urinary virus. Toulouse doesn't return to urinate in the closet to mark his territory; he's initially drawn by the scent of his urine to relieve himself there. Because the virus irritates the lining of his urinary tract and causes him to urinate more frequently, this behavior becomes strongly reinforced and firmly established. When he's downstairs, Toulouse can usually—but not always—make it to his litterbox; when he's upstairs, he can usually—but not always—make it to the closet. If he can't make it to these preferred locations, he squats wherever he happens to be when the urge overtakes him. However, because he produces only a few drops of urine which he scatters randomly throughout the large house, Hal doesn't notice the problem at first.

Most texts on basic cat care include discussions of feline urinary disease, and many veterinarians provide handouts on this subject to their clients, so we needn't dwell on the anatomical and physiological specifics of the disease here. However, always bear in mind that urination that occurs outside the litterbox might be in response to behavioral factors, but it might not. To be sure, gather as much information about where, when, and how the cat urinates, and take the cat in for a thorough veterinary examination.

When Hal rushes Toulouse to the vet's after learning about the urine-soaked closet, he can imagine only one solution to the problem: "I want this cat castrated immediately," he demands. "He's spraying everything!" However, the veterinarian's skilled fingers reveal a thickened and sensitive bladder, two sure signs of irritation. When he applies gentle pressure to the area, several drops of blood-tinged urine splash onto the stainless-steel table.

"How did you know he had an infection?" asks Hal, feeling simultaneously concerned for Toulouse's health and guilty about his misinterpretation of his pet's behavior. "I thought only neutered males can get urinary infections." The vet explains how cats with urinary tract infections initially adhere to one of two elimination patterns. Some retain urine because the greatest discomfort occurs when they urinate; these cats go to the box and urinate fairly large amounts and may cry out while doing so. Other times they may not make it to the box and will leave puddles of urine around the house. If the lining of the urinary tract becomes sufficiently irritated, as soon as the least bit of urine accumulates, the cat will attempt to void. In this case, the owner sees the cat in and out of the box constantly and may discover drops of urine all over the house.

For reasons unknown, cats with urinary tract infections often urinate in bathtubs and laundry baskets, on bedspreads, and on top of washers and dryers. Because this display more commonly occurs in otherwise healthy females and unblocked males, I personally suspect that these cats discover they can void pink urine in their litter until hell freezes over and no one will notice. On the other hand, a pool of pink urine in the tub or on clean laundry, the bed, or the washer generally elicits a prompt human response.

What about the way in which gender affects urinary tract infections? First, *all* cats, male or female, intact or neutered, can contract a urinary tract infection under the right conditions. However, as we noted in chapter 3, the long narrow male urethra can become obstructed by the debris associated with these infections much more easily. Second, I don't believe that castrated males block more often than toms; but I do believe that castrated cats that block are much more likely to be diagnosed and treated than intact toms.

Why? Let's consider the somewhat typical relationship between Denise and Interface. Denise allows Interface outdoors because when he reached sexual maturity, the stench of his urine permeated the entire house. She knows he roams, she knows he fights. Now suppose the shaded Persian contracts a urinary tract virus. He, too, will urinate more frequently just like Toulouse, but because he always relieves himself outdoors, Denise doesn't see this. One day shortly after Denise leaves for work, Interface blocks. He spends the entire day trying to urinate, he doesn't eat or drink. As toxic wastes accumulate in his system, he becomes depressed and crawls under Hal's garage. When he doesn't appear for dinner, Denise doesn't worry: "Must be out tomcatting around," she muses with a touch of pride.

She still experiences little concern when he doesn't appear the next morning, but she vows to search for him if he's not home when she returns from work. However, during the day the waste products build up so dramatically in the cat's body that he slips into a coma and dies at sunset. His owner searches the neighborhood, calls the animal shelter, places ads in the newspapers. At the end of a month, she gives up. "He must have gotten hit by a car or something," she tells Hal. Or perhaps, "He was such a beautiful cat, I bet somebody stole him."

The point is: If we judge the incidence of urinary tract infection strictly in terms of the cats diagnosed and treated, we might erroneously conclude that such infections rarely, if ever, occur in toms. However, they can and do, and owners of all males who relieve themselves outdoors should keep this possibility in mind. If you suspect a problem, get out the litter pan, keep the cat indoors, and observe him carefully. It could be a matter of life or death.

Fortunately, Toulouse's urinary tract infection responds well to medication. And by closing the door to the thoroughly cleaned bedroom, Hal counters Toulouse's natural temptation to return to the area long enough for the cat to forget about it.

REMOVING THE STRESS

Although delighted by his cat's rapid recovery, Hal must still deal with the now-daily clawing episodes and the increasing stress generated when Interface violates his property. After much hemming and hawing and several false starts, he finally broaches the subject to Denise.

"As soon as Toulouse finishes his medication and the vet's sure he's all right, I'm going to have him castrated," he announces cautiously. "The vet says that because the bladder infections are caused by viruses, stress can be a major factor, and I know that seeing Interface in our yard all the time really upsets him. We're hoping that neutering will at least reduce the stress associated with territoriality, if not eliminate it. But unless I can come up with another way to stop the scratching, I guess I'll have to have him declawed too."

The look on Denise's face tells Hal that his words have badly upset his friend. He suspected that such might be the case because she's always boasted about Interface's "wild independence," his macho fights, and breeding forays. She truly believes that the loving owner allows his or her cat to move without restriction whenever and wherever it chooses, and in its unaltered, natural state. When Hal proposes neutering and declawing as solutions to Toulouse's territorially related problems, her revulsion flows through her entire body until anger and frustration bubble to the surface. "How can you be so cruel and selfish!" she screams at Hal.

Over fifteen years ago an informal study was conducted to determine why people neuter or don't neuter their pets. Among those who strongly support spaying and castration as the only effective means of pet population control, most believed the cost of the surgery posed the greatest obstacle. After amassing and analyzing the data, the researchers announced a result that initially amazed everyone—until they thought about it. The study revealed that the choice to spay or castrate a pet rarely, if ever, hinges on the cost of the surgery. People who believe the procedure beneficial and indicative of responsible pet ownership do it, saving up or arranging time payments if necessary. People who don't believe it's right to neuter animals never have enough money, time, proper transportation, or an acceptable veterinary facility to accomplish the task.

Anyone who has ever stumbled into the middle of a discussion between the strongly proneuters and the antineuters knows that emotions run high, nobody changes anybody's mind, and soon nothing makes any sense at all. This occurs because each "side" sees the neutering issue as a symbol of intimate personal beliefs about sex and reproduction that may or may not have any relationship to the pet whatsoever. Fortunately and in spite of their passion, those who rigidly ascribe to either orientation constitute a minority. The majority of owners look at each cat and its relationship to them and their environment and objectively weigh the advantages and disadvantages of the surgery.

What are some of the advantages? In addition to population control, neutering decreases territorial behavior. Several studies by veterinarian-ethologist Ben Hart and his colleagues indicate that neutering definitely decreases fighting, spraying, and roaming in the majority of cats. Furthermore, one study indicates that castration confers these behavioral benefits even when performed on fully mature males. This is good news for those who prefer the more solid body build of the mature tom or believe that early castration and urinary tract problems are related, but still desire the positive behavioral effects of castration. On the other hand, no evidence supports the idea that allowing females to cycle before spaying increases their femininity in any way.

The negative effects commonly linked with neutering center around these animals' tendencies to gain weight and their lethargic behavior. As one of my professors pointed out, "Unless an owner can demonstrate how his or her cat goes into the kitchen, opens the bag or can, and puts the food into its dish, I consider weight increases following spaying or castration an owner problem—and responsibility. Responsible owners realize that neutered animals do require a lower caloric intake for several reasons. Often the surgery occurs just as the animals enter adulthood, when their normal growth rate slows considerably. Second, patrolling territory and mating consume a great deal of energy. If we feed the neutered animal who prefers to remain home with its owners the same amount we fed the free-roaming tom, we're bound to fatten up our cats.

What about lethargy? In an otherwise healthy cat, lethargic behavior depends on the eye of the beholder. Owners who considered their cats' active sex-related territorial behavior exhausting consider their neutered pets' behavior "mellow" and "like a normal cat." Many times owners who complain of postneutering lethargy shared minimal interactions with their pets in the first place. Perhaps they got the cat because they wanted a "self-keeper," an easy-to-care-for pet that was off "doing his (or her) thing" most of the time. Because they experienced so little contact with their pets, they enjoyed few if any shared activities. Consequently when neutering results in the cat giving up most of its purely feline activities, and with no human/feline interactions to fill this void, the pets do appear to do little more than sleep and eat. Unfortunately owners seldom recognize their own contribution to this phenomenon. If they compound the problem by feeding the cat the same amount, the resultant weight gain can further decrease activity, creating yet another vicious cycle.

Of course, substances as complex as the sex hormones do affect many aspects of physiology. Unfortunately the battle of the sexes and its accompanying passions and prejudices seems to extend even into the supposedly neutral realm of the research laboratory. For every study that says, "This hormone does x," there's another that says, "No, it does y," and yet another claiming it does both, which begets yet another study claiming it does neither. In truth, these substances are so complex and powerful (unprotected male workers in one pharmaceutical plant developed breasts and other feminine characteristics simply by handling female sex hormones) and we hold such strong personal beliefs about them that completely objective evaluation of their effects may never be possible.

Finally, no discussion about neutering can be complete without reference to its "unnaturalness." Only a fool would argue that removing ovaries and uterus or testicles under sterile conditions using anesthesia mimics a natural event in the wild feline's life. Again, this could launch us into one of those interminable passionate arguments wherein all participants project and counterproject their beliefs, with little concern for the cat's point of view. The answer to the question rests solely upon one's definition of *natural* or *normal,* which turns out to be as arbitrary and unique as each owner and cat. To Hal, the idea of a neutered house cat represents a totally normal—and loving—feline life-style, whereas Denise considers it appallingly restrictive. However, she would never consider *not* providing her pet with a fresh bowlful of gourmet canned food morning and evening, something the owner of the average barn cat considers the epitome of "human interference."

Never Say "Never"

Our opinions about castration and spaying usually reflect our quite human feelings about the subject, and because we create and accept these beliefs, only we can change them. About ten days after she storms out of Hal's yard in a fury over his intentions to castrate and perhaps even declaw Toulouse, Denise lounges comfortably in bed with Interface. As her hand passes over the thick fur on his shoulder, she thinks she feels a lump, but when her cat wheels and lashes out at her hand, she recoils in surprise. Then she becomes concerned: Could something be wrong with her pet? He has seemed a lot calmer lately, and he didn't even touch his breakfast. Even his presence on the bed, while delightful, is unusual. Normally, he'd have wolfed down his food and shot outdoors to investigate the neighborhood without giving her a backward glance.

Interface sits down carefully at the end of the bed and casts a baleful

THE BODY LANGUAGE AND EMOTION OF CATS

eye at his mistress. Suddenly he begins licking his shoulder with a vengeance, then he brings up a rear leg to scratch the same area more effectively. As a result, a thin layer of deteriorating skin in the area of the bite wounds easily gives way, and fetid pus and blood spray all over the bedclothes.

Now it's Denise's turn to rush panic-stricken to the veterinary clinic. Although the gaping hole means that the shoulder abscess will drain and heal well, other wounds on Interface's leg must be lanced. The veterinarian asks Denise whether she would like her cat castrated while he's anesthetized for wound treatment, a notion Denise vehemently rejects, but not without a touch of guilt. As the veterinarian escorts Denise out of the examination room, he reminds her to keep up with all of Interface's scheduled vaccinations: "Because he roams and fights, he's likely to encounter all kinds of viruses, and the stresses of territoriality and mating may lower his resistance." He also points out that Interface's squashed-nose Persian breeding could complicate any upper-respiratory infections, that his hunting habits bring him into contact with parasites as well as bats, which sometimes carry rabies in their area, and that sniffing urine marks deposited by other cats could expose him to urinary viruses. Contact with the urine and saliva of infected cats could also expose him to leukemia and infectious peritonitis.

Denise makes sure Interface receives all the proper veterinary care, but when he winds up at the vet's twice in the same month for anesthesia and treatment—once for more abscesses, once for a laceration—she begins to reevaluate her beliefs about neutering. Surely all that anesthesia isn't good for Interface. "It took him a lot longer to bounce back this last time, even though he was under the same amount of time," she tells Hal worriedly. (Recall what we said about cats and anesthesia in chapter 2; it's not unusual for cats to develop variable responses to repeated anesthesia.)

The day her once beautiful Persian returns home with his left eye swollen shut and the tip of his left pinna completely severed, Denise reverses her formerly adament opinions about castration without a moment's hesitation. When facing a choice between a matter of principle and the health and well-being of her beloved pet, like any loving owner Denise finds the choice a remarkably simple one to make.

THROWING DOWN THE GAUNTLET

If the discussion of neutering evokes passions, the subject of declawing arouses emotional fervor that puts religious fanatics to shame. Pick up any basic cat care book or magazine, and chances are you'll find something about declawing. While most aspire to be objective discussions, many invariably wind up leaning toward favoring or disfavoring the practice. Those articles offering a "total" view usually present several carefully worded paragraphs describing the disadvantages, balanced by an equal number describing the advantages of declawing.

One need only read ten or twenty of the hundreds of commentaries on the subject to realize that the bane or blessings of declawing, like spaying and castrating, lie solely in the heads of cat owners. For every person who argues that declawed cats can't defend themselves, another emerges with thousands of feet of film depicting his or her declawed cat successfully holding cats, dogs, and even threatening humans at bay. For every owner who claims the procedure hurts the cat unbearably (often evoking the anatomically unsound but highly emotional analogy relating declawing to the Oriental fingernail-yanking torture), another points out, "My cat came home from the vet's, immediately ate, used her box, and played with her toys as if nothing had happened." You say declawing turned the Smollins's cat into a vicious biter; your neighbors note it turned their once nasty Siamese into a perfect love. And around and around it goes.

Just as some owners long for some form of medical birth control that will free them from having to make a choice regarding surgical sterilization, so many owners seek a compromise for declawing when destructive clawing occurs. On the one hand, we want to eliminate the negative effects of the existing anatomy, physiology, and coordinating behavior; on the other hand, we don't want to shoulder the responsibility or guilt that may attend that. We noted that applying double-sided sticky tape to appealing scratching areas and discouraging the cat with water sprays may work if used consistently and initiated before the behavior becomes well established. Other owners find that keeping the claws clipped short will limit the damage. Because the nails are relatively thin and the blood vessel usually apparent, an owner can easily accomplish this task using standard human fingernail clippers or special veterinary ones—provided the cat accepts restraint. Also bear in mind that any advantages conferred by the claws in hunting, fighting, or climbing come from their sharp points; because of this, a cat with clipped claws incurs the same liabilities, if any, as a declawed one.

Another compromise to declawing reminds me of the two spouses who opt to buy a purple recliner because one wants a blue sofa and the other wants a red loveseat. This solution to the inappropriate clawing problem involves glueing wooden balls to the tips of the claws to prevent scratching. If we pursue the anthropomorphic orientation that convinces someone that this approach makes more sense than declawing, we can't help thinking that attempting to retract such claws would create discomfort, frustration, and annoyance, which, when added to the hassle of periodically replacing worn wooden balls, might well traumatize the cat as much or more than declawing. In this case, like the couple who wind up with the purple recliner neither likes, both owner and cat may find that this approach simply avoids the problem rather than resolves it in a way that meets with cat's or owner's needs.

Whether to declaw or not depends strictly on which approach will provide the most stable relationship between you and your particular cat in your particular environment. Because Hal is an interior designer and antiques dealer, chances are that Toulouse's clawing will ultimately lead Hal to experience anger, frustration, and perhaps even physically abuse his pet. Given these feelings, Hal may decide that he must either declaw Toulouse or get rid of him. If Toulouse only claws that one pair of drapes, Hal may decide to leave that window bare, or shut that room off. Or he may leave the tattered drapes hanging, cheerfully noting to visitors, "It's Toulouse's only vice, and I can live with it." Similarly, Denise may hold such strongly negative feelings about declawing that she willingly accepts any damage without allowing it to undermine her relationship with Interface.

How do you know you've made the right decision? In my experience people who evaluate their special situation with their particular cat objectively, then make a conscious choice to declaw or not to declaw rarely feel any need to defend their position. On the other hand, those who ignore early clawing episodes until the cat damages something valuable and then feel victimized, those who allow themselves to be pressured into declawing or not by others, or those who make impetuous decisions about declawing they later resent or regret will almost invariably pontificate their views on the subject. Having been on the receiving end of such tirades more than once, I usually get the feeling that these people hope to validate their position by getting me to agree with them. But if they're really so sure that their choice represents the best for themselves and their cats, what I think shouldn't matter at all. Therefore such passionate proclamations usually lead me to question the owner's position rather than to support it.

If you catch yourself spouting passionate and rigid opinions about declawing or any other aspect of your relationship with your cat, you may want to take a hint from Shakespeare and figure out why you're protesting so much. Perhaps changes in your relationship with your cat are forcing you to examine a new orientation you'd rather ignore. For your cat's and your own sake, try to evaluate all the alternatives objectively and don't be afraid to change your orientation if it means a better relationship for you and your pet.

BURSTING THE TERRITORIAL BUBBLE
Although four days elapse before the Clarks actually do something about Reject, they know the instant they receive the letter from the adoption agency that there's no way cat, toddler, and adults can live

The Body Language and Emotion of Cats

quality lives together. Reject would live in a constant state of fear because the Clarks could not possibly protect her from the explosion of sights, sounds, smells, and touches associated with an active child. Nor could they possibly expect a three-year-old to adhere to the almost monastic way of life they created to avoid upsetting the cat. And finally, they know they can't trust Reject. Already she's lashed out at them countless times with teeth and claws, responding to threats beyond their comprehension. Even when they expected such an attack, it still startled (and hurt!) them; a toddler would surely not fare so well. With heavy hearts and little hope they set about trying to find a suitable home for their cat.

Because of the crucial development that occurs during kittenhood, many veterinarians, behaviorists, and breeders would have strongly advised the Clarks not to take Reject in the first place. However, for every expert advancing such cold, logical recommendations, there exists a potential owner who knows that love can and will conquer all in the long run. Most experts will agree that, given enough time, a stable quality relationship could evolve from the most inauspicious beginnings. However, they also agree that it usually takes much longer and requires a great deal more patience than the idealistic feline saviors anticipate. When results don't measure up to expectations, or changes in the environment convert the challenge to a burden, owners often succumb to our old nemeses: anger, frustration, and guilt.

If you ever face a situation similar to the Clarks', first eliminate all thoughts that begin with "I should have . . ." or "If only I'd . . ." It doesn't matter what you could have or should have done in the past; all that matters is what you're doing and intend to do now. Poorly socialized animals with unstable territorial responses require delicate handling. If you can't provide that with the full understanding that you may have to live with certain problem behaviors for the animal's entire life, it would be far more humane to terminate the relationship, either by placing the cat in a more suitable environment or by having it euthanized.

While I oppose hasty or poorly considered euthanasia, I even more strongly oppose dumping a poorly socialized animal into yet another difficult environment simply to avoid confronting this option. With only ten days to effect any transition, the Clarks feel that any new environment would subject their pet to unconscionable psychological trauma. Sobbing, but comforted by the knowledge that they made the right choice under the circumstances, the Clarks hold and soothe the frightened cat as the veterinarian puts her to sleep.

Suppose that rather than discovering they're about to become instant parents, the Clarks discover that Ms. Clark is two months pregnant. In this situation, they continue their arduous socialization while at the same time locating a person who has both the expertise and environment to give Reject a good home. During the next seven months, Reject gets acquainted with her new owner and then her new home. Again, time offers the key to achieving a successful transition. Kittens seem to accumulate a tremendous amount of knowledge in a remarkably short period of time, but unlearning counterproductive knowledge often takes much longer. It's as though the lessons of infancy and kittenhood get stamped indelibly on the brain, while those of adolescence and adulthood get printed in water-soluble ink.

Whenever you find yourself in the position of potential savior to a "problem" animal, recall the Oriental tradition whereby the saved individual owes his or her life to the savior. Although such a notion might boost the ego in the beginning, many who choose this route discover that being another's sole reason for existence, another's whole world, can turn into a crushing burden. The poorly socialized cat whose owner constitutes its whole world often displays such a wide variety of negative territorial behaviors that the unknowledgeable savior winds up wondering why he or she didn't learn more about what lay ahead before undertaking such a task.

FINDING THE RIGHT PLACE
When we put all this information together, we become aware of yet another troubling paradox. On the one hand, a cat's personal and physical space requirements depend to such an extent on early experiences

that they resist even the most heroic efforts to change them. On the other hand, they can be so highly flexible that a cat may unaccountably accept one person in its space and viciously reject another. How can we and our cats hope to harness such mutually exclusive and powerful forces?

As we discovered in our discussion of nocturnal behavior, the solution lies not in accepting one orientation or the other, but rather in recognizing that the territorial response results from the balanced interplay of the two. If we understand the critical role early development plays in the cat's ability to adapt to a particular physical environment, we'll pay more attention to the physical surroundings when we select a kitten. If our adult cats strike us as too aloof or unfriendly, we can reconstruct their early experiences for clues to help us become more accepting and tolerant of asocial behaviors.

Our awareness of early experience in the development of territoriality also keeps us from falling victim to another particularly insidious owner belief. A surprising number of people choose to believe that the cat or kitten who manifests a need for more personal space (that is, it reflects the more asocial adult feline nature rather than the more human social orientation) must have been beaten or abused. This evaluation most commonly befalls animals obtained from shelters or taken in as strays. Although we can't deny the existence of that flawed human minority who do deliberately abuse animals, most asocial feline behavior results from other, emotionally neutral environmental factors. Remember what we said about feline perception of vertical and horizontal objects, food preferences, sound and scent habituation? A kitten who spends several weeks in a cage during critical developmental periods will assuredly construct a different worldview than one who spends that same period exploring a suburban split-level. The stray who wanders into your condominium from the farm may need more personal space than the lost cat raised by a devoted family.

The assumption that a kitten or cat requiring more personal space was abused gets any relationship off to a shaky start because, more often than not, it leads owners to rationalize their pets' negative behavior rather than come to grips with it. When the Clarks stop entertaining because "Reject was abused and doesn't like strangers," they shift the responsibility for this change in their life-style from themselves to the cat. This works well as long as they don't want to entertain in their home. However, if six months later they decide they'd like to receive guests again, this abrogation of responsibility precipitates two negative results. First, they resent their cat for putting them in an uncomfortable position. Second, they still suffer the consequences of the negative behavior. And because they have permitted the behavior to persist six months longer, it will be that much more difficult to alter.

Understanding the interaction of personal and physical space and asocial and social orientations also enables us to see those blue-eyed, people-loving members of the feline population and their relatives as exquisite examples of nature's order rather than as exceptions to the natural rule. Given such cats' vulnerability in the physical realm where other felines possess protective coloration and the full range of sensory ability, these individuals, whether by chance or by choice, evolved territorial concepts related more to human presence than to physical space. And because they depended on people for their food much as a kitten depends upon its mother, their dependency continued into adulthood until the social orientation became a permanent rather than a transitory aspect of their personalities.

This phenomenon extends beyond the modern spin-offs of those first Siamese mutations. As the bond between human and cat evolved over the centuries, genetic and environmental factors produced cats who redefined their territorial instinct in terms of people rather than in terms of fixed physical spaces providing food, security, and mating opportunities. Because human interaction can offer these same benefits (including providing the perpetuation of the surrogate queen/kitten relationship, which along with neutering precludes the desire to mate), we can understand how easily such a transition could occur.

When we consider the wide range of territorial expressions, we can

THE BODY LANGUAGE AND EMOTION OF CATS

appreciate why some cats can readily adapt to other cats in their environment while others find the presence of other felines intolerable. We can also see why cats who define their territory in terms of their owners may resist changes that are different from those who claim a physical space. Interface finds only the presence of other cats in his territory intolerable; Toulouse reacts negatively to the presence of *any* animal and some people in his house; and Reject considers anyone and anything but the Clarks, her food and water bowls, and a very few toys a territorial violation. Only by recognizing the individual cat's orientation can we predict whether and how that cat will react to intruders in the household.

Although each cat expresses its own unique form of territoriality, some general trends do exist. The "average" indoor/outdoor cat expresses a combination territorial affect, claiming both owners and physical space as part of what it needs to fulfill its needs. In this case, the bond to either human or physical place isn't so strong that new owners or a new physical location would create anything beyond a temporary period of readjustment.

Sometimes cats who claim a complex territory, such as a city street, will maintain two or three different "homes," often without each owner being aware of the other's existence. In such a way these cats simultaneously fulfill their needs for human companionship and preserve their physical territories in a most ingenious and economic manner.

At either extreme of the territorial spectrum we have those cats that bond strongly with *either* a physical territory *or* a person. This explains why some cats exhibit great reluctance to move to a new home with their owners, whereas others insist on sticking with their owners wherever they go. Studying cats possessing these two strong orientations can teach us a great deal about how cats attempt to maintain their territory, be it an actual physical space or a person.

Mounting evidence suggests that an animal's awareness of its fixed physical territory includes knowledge of the skies above it. Consequently when removed from their home grounds, these animals can eventually find their way back via celestial navigation. Although such ability seems extraordinary to humans, migrating animals and birds routinely use this technique. However, one drawback to the celestial-navigation theory is that it would seem to require a degree of visual acuity in limited light beyond that attributed to cats. Finding and following the Big Dipper on a starry night would seem an impossible task for an animal with limited-detail vision. Perhaps cats respond not to the seen but to the unseen electromagnetic forces. If so, cats may actually orient themselves to the magnetic north pole rather than the stars.

Does that sound too preposterous? Some recent research indicates that at least some kinds of whales follow specific electromagnetic "paths" in the ocean, much as cars follow a highway. The adherence to such trails can be so strong that some individuals may follow them even though they lead straight to dry land and certain death for the beached whale unless help arrives. If such clear and strong messages can be communicated between sea mammals and electromagnetic currents within the ocean, we can't ignore the possibility that similar paths may steer land mammals too.

Although celestial or electromagnetic navigation could explain how cats possessing a strong sense of physical space find their way home, it doesn't explain how cats like Sugar can follow a family fifteen hundred miles from Oklahoma to a new home in California. Or does it? For insight into this form of nongeographic psychological territoriality we need to explore that gray area between psychology and parapsychology, the normal and the paranormal. The late Dr. Joseph Rhine, well-known director of the prestigious Parapsychology Laboratory at Duke University, and his colleagues devoted years to the scientific collection of data about such journeys. In order to be considered a valid case in point, the cat must possess some distinctive physical identifying mark, it must be alive and available for examination, and there must be eye-witness verification of the owner's claims. Despite these stringent criteria, Dr. Rhine and his colleagues amassed impressive evidence of what they called *psi-trailing*, the ability of an animal to trail its owner to a previously unknown location. When cream-colored Persian Sugar showed up in California, his deformed left hip joint convinced his owners and researchers that this was truly Sugar and not a cat of similar color, conformation, and temperament. And as if to silence skeptics, one spunky feline trailed his owner, a veterinarian, from New York to California. While the owner could perhaps overlook the strikingly similar appearance of the cat that showed up in California and the one left behind in New York, and even the fact that the "new" cat immediately took possession of the "old" cat's favorite chair, what are the odds both cats would also have enlarged fourth caudal vertebrae?

Unraveling the mysteries of how cats reclaim their "human territories" via psi-trailing can provide cat owners with a fascinating and often enlightening pastime. Those drawn toward a religious explanation may see psi-trailing as proof of God: "Are not two sparrows sold for a penny? And not one of them will fall to the ground without your Father's will" (Matthew 10:29). Surely if the Creator watches over sparrows, He takes care of cats separated from their loving owners as well. Others may equate psi-trailing with Kipling's words reminding us that cats dwell in a different world we do not understand: "The Cat. He walked by himself and all places were alike to him." Still others might view the reuniting of owner and cat as positive proof of Virgil's observation that "Love conquers all things."

While I dearly love the images evoked by these quotations, the logician within me responds to Aesop's two-thousand-year-old question: "Who shall bell the cat?" Among the first experiments performed in outer space was one designed to test the validity of a different kind of bell, Bell's Theorem. Basically Bell's Theorem proposed that all electrons function as pairs, with each electron always spinning in a direction opposite to that of the other; furthermore, if the spin of one electron changes, the other automatically "knows" and alters its direction in order to retain the opposite orientation. Physicist Bell postulated that such a close and sensitive relationship and bond at this level must exist

Territoriality: To Have and To Hold

to explain many of the events we take for granted in our natural world. Sure enough, when one electron was sent into space while its partner remained on earth, theory became reality. As soon as the spin of one electron was reversed, its partner immediately made the necessary adjustments to maintain the proper relationship.

Such microscopic and inanimate awareness of relative position and change applies to closely related individuals independent of separation by space. You and I and our cats consist of molecules and atoms, our atoms of electrons and other particles. Some studies of maternal instinct indicate that at least some mothers seem to know instantly when some (usually negative) change befalls a loved one. Although human logic and perception can't begin to explain such occurrences, our pets do seem capable of similar "extrasensory" feats. As we develop more sophisticated equipment, we may discover that some phenomena currently beyond our comprehension that appear almost magical or mystical simply lie beyond our sensory capacity. We can't hear in the ultrasonic range, but we now accept that such sounds exist and affect the lives of animals who can hear them. Now that we have more sensitive equipment, we realize that the "psychic" bond between queen and kitten relies not on a completely unknown, incomprehensible mechanism but rather on an extension and variation of our own familiar senses into realms beyond our unassisted perception.

Lacking a total understanding of every aspect of territoriality, we may come to see our relationships with our cats the same way Arthur Miller saw Willie Loman's relationships with the world: "way out there in the blue, riding on little more than a smile and a shoeshine." When things go wrong and boundaries change, the personal and physical space may tremble a bit, but if the relationship revolves around a solid bond, the balance will be restored because it, too, goes with the territory.

Regardless of how much we try to isolate nocturnal, territorial, or predatory displays, maternal body language and sexuality weave through

THE BODY LANGUAGE AND EMOTION OF CATS

all feline behaviors, taking turns so unexpected that few can claim to comprehend them fully. When Father Paul's Siamese, Aquinas, sucks a hole in the priest's woolen ceremonial cloak, only divine and feline grace and agility keep the cat from being hit by Father Paul's flying slipper. Can such disgusting behavior possibly be related to feline sex? Is it possible that a perfectly normal kittenhood body language display can create such unholy havoc in the relationship between the adult cat and its owner?

7

FELINE SEXUAL AND MATING BEHAVIOR: TOO CLOSE FOR COMFORT

*T*HE Garcias' friends so admire the extraordinary personalities of the couple's Siamese cats that they all cheer when the Garcias decide to breed the pair. After ascertaining the perfect health of both cats and bringing them up to date on all their vaccinations, the Garcias settle back and wait for nature to take its course. It does, and following a brief but noisy romantic interlude, the Garcias and their cats resume their normal routine.

The day the queen gives birth proves memorable for two reasons. First, the birth, like the mating, proceeds flawlessly to its completion, yielding five gorgeous, healthy kittens. Second, Steve Garcia's long-awaited promotion finally becomes a reality; in less than three months he and his family will move to corporate headquarters in Miami. Although they'll miss their New York friends, they're thrilled to be returning to their hometown.

In light of the impending move, the Garcias want to find homes for the kittens as soon as possible. When Steve's brother, Paul, a priest from New Rochelle visits, he admits that he's toyed with the idea of getting a parish cat. Not surprisingly, he selects the smallest kitten in the litter. On the way home, Paul dubs the kitten Aquinas and begins forming a tie that will eventually bind him to his cat in unexpected ways. Because of Aquinas's small size, Paul carries him around in the pocket of his baggy old cardigan sweater. Although his brother assured

THE BODY LANGUAGE AND EMOTION OF CATS

him that the kitten was weaned, Aquinas seems hesitant about eating. However, he appears healthy, playful, and normal in every other way.

When the kitten outgrows the pockets of his sweater, Father Paul feels relieved. "A good thing, too, young fellow," he tells Aquinas, wiggling his fingers through several holes in the garment. "You've turned it into Swiss cheese!" Later Paul smiles as he watches Aquinas fashion himself a nest in the sweater, then knead and purr himself to sleep.

Several months later Father Paul spies Aquinas squeezing through the narrowly opened hall closet he uses for storage. "Lucky I saw you or I could have shut you in there," he scolds his cat good-naturedly. Glancing in the closet, he notices a gaping hole in an overcoat. "Holy cow!" he exclaims angrily. "I thought you'd outgrown that bad habit." He forgets about the incident until the day he leaves his woolen cere-monial robe lying on the couch. As he enters the living room to retrieve the garment several hours later, he discovers Aquinas con-tentedly curled in the center of his woolen nest, sucking on the fabric, blissfully purring and padding. A few impure thoughts flash through the priest's mind as he deposits the cat outdoors.

Nowhere does a desire to view the world of cats in terms of abso-lute reproducible results rather than variable relationships generate more problems than in the study of sexual and maternal displays. Fur-thermore, we find our understanding of these areas of normal behavior severely hampered by three seemingly unworkable contradictions. First, although sexuality always holds great fascination for us, our culture and society also imbue it with untold inhibitions and taboos. I suspect that the book *Everything You Wanted to Know About Sex but Were Afraid to Ask* captured the public fancy because the title struck a most respon-sive chord: We *do* want to know, but something makes us afraid to ask.

Second, we often expect the most specific data about a phenomenon whose beauty arises from its incredible nonspecificity and flexibility. Why do males do this and females do that? Is this behavior normal or not? The answer, of course, varies from individual to individual.

The third, and perhaps most damaging contradiction arises when we erroneously equate *different* with *wrong*, applying our absolute ideas about what constitutes normal male or female behavior and judging others accordingly. Furthermore, because we link maternal behaviors so closely to our ideas of femaleness, our beliefs about what constitutes good maternal behavior tend to follow equally rigid and paradoxical guidelines.

Our beliefs about sex and parenthood precipitate far-reaching effects on our relationships with our pets, so we need either to accept or to resolve these contradictions before we can objectively analyze feline sexual and maternal displays and solve their related problems. Not to do so amounts to trying to play bridge when you don't know the rules. Although you can eventually get the hang of the game by observing others and applying trial and error, an awareness of the principles underlying it could save you a lot of time and frustration and increase your enjoyment of the experience immensely.

Resolving the dilemma created when people simultaneously want and don't want to know about sexuality eluded Freud and his followers and certainly lies beyond the scope of a book on feline body language and emotion. However, we should bear in mind that resolving this or any question invariably hinges on our willingness to make choices. We must either choose to know or choose to remain ignorant. Since either response may be the "right" one depending on the circumstances, the important fact isn't so much which choice you make but that you recognize it as a choice. In general, however, we may say that owners who choose ignorance regarding sex-related *problem* behavior invite numerous assaults on the bond between human and feline.

For example, suppose that Father Paul feels guilty for evicting Aquinas and atones by presenting his pet with a brand-new catnip mouse. Most adults would intuitively recognize Aquinas's response to the toy as sexual in nature, even if they've never seen cats mate; yet rarely do they openly admit to their awareness of this association. Father Paul, like so many of us, enjoys watching his cat frolic with the catnip toy but doesn't delve into the actual nature of the behavior.

There's nothing wrong with this approach so long as we find the behavior acceptable. However, when any feline sex-related behavior bothers us and therefore threatens to undermine our relationship with our cats, a knowledge of the why and how of that display becomes critical to any effective solution. If our own feelings of embarrassment or disgust preclude an objective analysis of the problem and inhibits any discussion with those professionals who may be able to help (veterinarians, behaviorists, breeders), we hang on to our personal orientations at the expense of a solid bond with our pets.

When it comes to the bond between human and feline, this choice to remain ignorant almost always functions negatively because we still pass judgment on behaviors we find irritating, even if we choose not to understand how or why these occur. So many times, those who possess the least knowledge pass the harshest judgments. This results in the Foghorn Principle: The less some people can see, the more noise they make. Woe to the cats whose foghorn owners condemn or even punish behaviors about which they know little or nothing, for those poor creatures often become helplessly enmeshed in a vicious cycle. Father Paul sets ever more elaborate traps to catch Aquinas sucking and punishes the Siamese more and more severely every time he catches his cat displaying the behavior. Although the owner's responses escalate both physically and emotionally, they may do little to stop the behavior and a great deal to undermine the relationship.

Consequently, there are only two valid resolutions of the knowledge/ignorance dilemma that can maintain a stable bond, and we must make a choice between them:

- We can opt to remain ignorant but *accept* the behavior, regardless of the form it takes.
- We can choose to learn about the behavior, putting aside any negative emotions (embarrassment, disgust) that acquiring that knowledge may create.

Feline Sexual and Mating Behavior: Too Close for Comfort

Father Paul can throw up his hands and proclaim, "Aquinas is destined to suck holes in things the rest of his life, and that's the way it is." Or he can say, "I really like Aquinas and want to keep him, but I can't tolerate these holes in my belongings. Even if the vet thinks I'm weird and my cat's a nut case, I've got to find out more about this behavior."

We've encountered the paradox of specificity versus generality throughout our discussion of feline behavior. Because sexuality and maternal displays involve the relationship between two or more individuals (or an individual and an object as in Aquinas's case), we must replace absolutes with probabilities. Although we can state that cats generally respond to their owners as kittens to queens, each queen exhibits her own unique form of interaction with her kittens and vice versa. These highly variable, individually unique responses then affect the quality of the human/feline relationship. Maggie doesn't always repond to my petting her the same way, but there are elements of her response that occur whenever *anyone* pets her, as well as those she *only* displays toward me. In other words, her response to petting contains both highly relative and highly specific elements.

The ability to distinguish these differences proves particularly helpful when analyzing problem behavior. Compare these two statements Father Paul might make about Aquinas's behavior:

"Aquinas sucks holes in everything!"

"Aquinas sucks holes in my woolen clothing."

Which one tells us the most about the true nature of the display?

Finally, we need to eliminate the all too prevalent and erroneous link we make between *different* and *wrong*. The word *different* springs from an objective analysis, whereas *wrong* results from an emotional one. This distinction helps us properly evaluate any behavior, but it makes a huge difference when we analyze sexual displays which, by their very nature, tend to arouse a wide range of very potent human emotions. Moreover, our highly emotional ideas about what constitutes right and wrong sex-related behavior often spring more from ignorance than knowledge. Whether these quasi-definitions are projections of our beliefs about normal human behavior or how we think cats *should* act, they block our ability to evaluate the behavior objectively and seek out additional information if necessary.

People still routinely observe cats mating and gasp, "My God, he's killing her!" They encounter a female in heat and interpret her normal body language expressions as evidence of severe brain disease. Their adult cat begins kneading when they stroke it and they angrily fling the animal to the floor screaming, "Ungrateful wretch, trying to scratch me when I'm being so nice to you!" None of these people realize that all they're doing is generating a lot of negative emotion that does nothing to increase their knowledge but a great deal to undermine their relationship with their pet.

Compare these responses to that of the person who observes cats mating and says, "My, they certainly behave differently from dogs, don't they!" and who moves naturally toward more specific questions and observations that lead to new knowledge and understanding. Those who calmly observe females in heat quickly realize these animals are neither in pain nor demented, but rather exhibiting one of nature's most fascinating behavioral displays. Those who resist flinging the kneading cat to the floor and wonder, "What's he doing? I never saw this behavior before," almost invariably congratulate themselves for their objective observation, "Why he's acting just like a contented kitten nursing!"

VIVA LA DIFFÉRENCE!
When cats enter the world, they possess the potential to be both very much alike and very different. All kittens share some very specific displays, such as greeting, grooming, playing, and stalking behaviors. They also exhibit displays common only to members of one sex. Females go into heat and engage in maternal behaviors; males mount, spray urine, and generally react more aggressively toward other males. However, these sexually related traits defy clearly marked boundaries.

Feline Sexual and Mating Behavior: Too Close for Comfort

Female cats, especially those with kittens, may spray urine to mark their territories with an enthusiasm and skill worthy of any tom; and all veterinarians have heard tales of male cats whose behavior toward a litter of kittens in the same household could only be described as maternal.

In this era of feminism, liberation, and sexual revolutions and counterrevolutions, the idea that males and females might share more traits than not could start a free-for-all. However, an unemotional approach to sexuality reveals some exquisite details often obscured by the passionate dust raised by the battle of the sexes. Consider an elegantly simple truth: To be female is not to be male. In other words, females are females not so much because they secrete special substances or hormones that make them female, but because they lack those that make them male. From an evolutionary point of view, why would an animal evolve two separate systems when it could achieve the same result by making the second system a variation on the first?

For example, suppose you wish to convey two different messages, x and y, to a friend using a flashlight. You could decide that a green light means x and a red one y, but this means that you will need two different colored lights or different colored shades or filters to produce the two colors. Why not simply tell your friend, "If the light's on, it means x, and if it's off, it means y." Because this solution requires only one light and no shades or filters, it not only requires less energy, it provides fewer opportunities for miscommunication.

However, being male or female consists of more than the relative presence or absence of certain hormones and anatomical parts. We know, for example, that sometime shortly before or after birth males experience a transitory surge of testosterone (one of the male sex hormones), which scientists believe sensitizes certain organs, including the brain, to respond to that hormone at a later time. When that hormone is again secreted during puberty, these sensitized organs develop in the characteristically male fashion. The lack of testosterone during this critical early period orients the individual toward more feminine development.

Although scientists haven't uncovered the exact nature of this mechanism, we do know that differences in the brain rather than the reproductive organs themselves determine sexual behavior.

Several groups of investigators have demonstrated this phenomenon by injecting adult cats with opposite-sex hormones. Castrated males given estrogen, a female sex hormone, don't behave like females; nor do spayed females given testosterone display male sexual behaviors. The tendency to respond in accord with the "mind-set" rather than circulating hormones is so great that females given testosterone may respond as though they had been given estrogen. So while we humans may have some pretty definite ideas about what comprises maleness or femaleness, physiologically the differences aren't nearly so apparent.

THE LOOK OF LOVE

What makes for "good" cat sex? For dogs, a good "tie" or mating physically links the partners for half an hour to forty-five minutes, whereas that between cats lasts only a few minutes. Some owners find it a bit disquieting that the cat's participation in a longer period of foreplay followed by intense physical interaction is quite unlike the dog's—but not unlike the human's. Cat sex noisily presents us with the kind of question we'd rather ignore: Does the display intrigue (upset, repulse, embarrass) us because it's so alien or because it's so familiar?

In order to understand how feline sexual behavior can affect our relationships with our cats, we must first recognize the nature of the act itself. Bear in mind that reproduction constitutes a primary species drive. Without reproduction the species cannot survive, regardless of how efficiently its members establish and protect their territories or pursue and capture prey. Therefore, even though certain behaviors associated with mating may strike us as unnecessary or even bizarre, we must accept the fact that they evolved over the centuries to benefit the species and ensure its survival.

During our discussion of territoriality we noted that receptive females announce their presence via mating calls and specific hormones

Feline Sexual and Mating Behavior: Too Close for Comfort

they secrete in their urine. Males respond to the signals with their own beckoning and identifying vocal and scent cues. Let's follow a pair of cats from the instant each becomes aware of the other's presence.

If either cat sniffs the urine of a member of the opposite sex, that cat pulls its upper lips back to expose the duct of the vomeronasal gland and block the nostrils (the flehmen reaction described in chapter 3), which facilitates processing of sex-related scents. These odors are so specific and potent that a receptive female may assume the mating posture immediately, even if the male isn't physically present. Shifting most of her weight to her front legs, she crouches and elevates her rear end while holding her tail to one side and treading with alternate steps on her hind legs.

Some observers have noticed that domestic cats will, for reasons unknown, display this same behavior toward their owners or other people. My own experience and that of several clients supports this observation. Sometimes the presence of an adult human male appears to trigger or intensify heat-related displays in female cats. In these cases the cat, typically a house cat, assumes the mating posture and commences her plaintive cries the instant a man enters the scene, even though she has given no indication to any women in the household that she was in heat. The cat often pursues the man—who may not even like cats—throughout the house. The man who tries to escape to bedroom or bathroom encounters the feline seductress, body pressed tightly against the door, treading, trembling, and yowling the anguish of her unrequited love. Such persistent adoration can be tiring, if not absurdly funny. In my experience, this crossover practice more often occurs among Siamese and other strongly people-oriented breeds or individuals. Considering the willingness of such animals to form close attachments to people and respond to them as cats from an early age, this behavior shouldn't surprise us. Because these cats see their owners as adult cats, why wouldn't young females automatically seek out human males as potential mates when they become sexually mature?

In the actual presence of a male cat the female's mating response is even more powerfully triggered. In this case, flehmen reactions and genital examinations (to ascertain receptivity and suitability) succeed nose-to-nose general-greeting displays. If the cats find each other physiologically and psychologically compatible, serious mating ensues. The male grasps the female's neck in a manner quite similar to that used by queens to immobilize their kittens. Positioning himself a bit forward on the female's back, he, too, begins treading with his hind legs, gradually sliding backward until he aligns his penis with the female's vaginal opening.

We don't know for sure why cats tread during sex, but anatomy does offer a few clues. Recall that the relaxed tom's penis faces posteriorly. Consequently, if he were to position himself directly behind the female immediately, he would not be physically able to stimulate an erection. Perhaps the combination of the slide toward the female's rear end, the rhythmic treading of both partners, and the numerous other sensory cues converging in those ecstatic moments serve to get the job done. If so, then the barbs on the penis may very well function more to prepare the tom to mate by increasing stimulation of the penis than facilitate the mating act itself.

As intercourse proceeds, the female continues treading while the male commences pelvic thrusting. These thrusts increase in frequency and depth until ejaculation followed by one last deep thrust occurs.

After the final thrust both cats remain motionless for a few seconds. Interestingly, while the male's excitement level now decreases, the female's increases. Her pupils dilate, and she begins to pull away, emitting a piercing cry. Once separated, she turns and lashes out at the male, then begins licking her genitalia. Finally, she performs what behaviorists call the afterreaction: She rolls and rubs herself on the floor or ground with obvious delight. Perhaps this results from physiological and hormonal changes associated with ovulation that occur as a result of mating and is not independent of it, as in other species, such as humans and canines.

Recall our discussion of circadian rhythms and the daily and seasonal fluctuations of hormones. During heat, the amount of the female sex hormone, estrogen, gradually increases to become the dominant hormone. This delicate balance between estrogen and the other hormones, however, changes dramatically when ovulation occurs. Suddenly estrogen production drops and another female hormone, progesterone, dominates the scene.

Although we primarily think of progesterone as a sex hormone, it also wields formidable behavioral power. In low doses it inhibits both the male and the female sexual responses. As the level increases, it creates a sense of tranquillity or well-being and euphoria; at even higher levels, it functions as a general anesthetic. Because increased progesterone levels accompany ovulation, and given the fact that cats are reflex ovulators, a sudden burst of progesterone may well explain the after-reaction. First we see the total extinguishing of the sexual response, which manifests as repulsion of the previously beloved, followed by a euphoric state as the progesterone levels continue to rise.

So what the Garcias may view as a loving courtship followed by their female's vicious attack on the male and her delirious delight in *his* confusion comprises a perfectly normal physiological response. By viewing it in this light, the owners can avoid needless embarrassment or, worse, negative evaluations of their male's sexual display as brutal and the female's as victimization followed by (divine) retribution. Compare the objective, unemotional view to that of one client who, following a perfectly normal mating between her cat and her neighbor's tom, refused to speak to her neighbor because "She just stood there and let her cat rape my baby!" Needless to say, this woman's view of feline sex affected her relationship with both cats and humans!

MATERNAL BEHAVIORS: BACK TO BASICS

Relatively speaking, asocial cats experience fewer problems giving birth than most domestic species. Barn cats routinely disappear to give birth, then raise their young to the point where they can fend for themselves before presenting them to their owners. While this may perplex and

THE BODY LANGUAGE AND EMOTION OF CATS

disappoint social humans, who love to share in the miraculous process, it suits the feline temperament perfectly. At least, it did in the past, when the cat's solitary orientation helped weed out the reproductively and maternally unfit. Queens who couldn't give birth unassisted perished; although those with poor maternal instincts might survive, their offspring wouldn't, thus ensuring that the trait would not be perpetuated in the gene pool.

In order to understand some of the behaviors that attend maternal behaviors, let's briefly describe the events that accompanied Aquinas's entry into the world and influenced his development. First, shortly before the actual delivery began, the amount of progesterone circulating in the queen's bloodstream increased. We know that sufficient levels of this hormone produce a sense of tranquillity and anesthesia, and some physiologists believe that it thus enables queens to give birth with minimal stress and discomfort. As the fetuses move down the birth canal, they create pressure, which in turn causes fluid to leak through the vaginal opening. The presence of this fluid stimulates the queen to lick the area, a process she continues until a kitten is not only delivered but is also free of the fetal membranes and afterbirth. Not only does the licking help dry the kitten but as already noted, the roughness of the queen's tongue stimulates most kittens to lodge vigorous vocal complaints, a behavior that serves to clear the respiratory passages of any fluid. Most queens take their licking so seriously that they take the full thirty-to-sixty-minute (average) interval between kittens to accomplish this task.

Because the Garcias knew that queens licked more in response to the fluids than to the kittens themselves, they were careful to line the queen's box with a thick layer of absorbent towels several days before her due date. That way they were able to guarantee that the newborn would be wetter than the bedding and therefore likely to warrant most of the queen's attention. Although not a common occurrence, most veterinarians have shared the anguish of novice owners of inexperienced queens whose kittens die of suffocation in the birth sack or of exposure because the queen spent more time licking herself and her bedding than her kittens.

Soon after birth, Aquinas and his littermates found their way to the queen's nipples, which she conveniently exposed once she had delivered the whole litter. Although at one time scientists believed that the kittens were drawn to the teats by smell, current theory holds that the mother's warmth provides the strongest incentive for the young to move toward her. This explanation seems the more logical because newborns possess only a limited capacity to maintain their body temperature, and becoming chilled more often leads to fatal complications than lack of nourishment.

Unlike many social litter producers, whose young immediately begin establishing a pack structure by staking out a particular teat for the entire nursing period, kittens may or may not nurse at a particular location. Regardless of the location, all kittens practice the same nursing technique, flaring the toes of their front feet and pressing against or kneading the mammary gland to enhance milk flow to the nipple. As we would expect, behaviors established during this period become quickly and firmly imprinted; this explains why people who begin bottle-feeding orphan kittens often have trouble transferring this responsibility to a willing feline foster mother later.

Following nursing, the queen grooms the kittens, concentrating primarily on the area around their rear ends. This stimulates the young to urinate and defecate; the queen then licks up the wastes and swallows them, thereby keeping both the kittens and the bedding clean. Not only does such an efficient process prevent the spread of disease among the young, it also removes scents that could attract predators to the nest site.

The queen usually initiates nursing during the first three weeks, confining it to the nest area, which she keeps thoroughly clean. Around the third week, the kittens' eyes and ears function well enough for them to venture out of the nest and begin exploring the larger world. Prior to this time maternal retrieving behavior serves to keep the litter contained. Queens instinctively retrieve kittens in response to their cries, a response that peaks when the young are about one week of age, then diminishes as the kittens' eyes and ears open and they develop the ability to maintain their own body temperature. Considering their extreme vulnerability during this period, the queen's display can mean the difference between life and death for the kittens.

Once wild kittens begin their explorations outside the nest, another maternal instinct comes into play. Instead of retrieving her young to a fixed location, the queen will more likely move the entire litter to a new location, particularly during the third through fifth weeks. The survival benefits conferred by this behavior in the wild are obvious. As the kittens begin moving and depositing their wastes further and further from the nest, the chances of predators locating them greatly increase. By relocating the home base whenever she senses that environmental changes might threaten her young, the queen enhances their chances of survival.

Although we can readily equate the maternal behaviors displayed by the Garcias' Siamese with those of the wild feline, domestication provides its own special variations. Because of her strong relationship with her owners, the Siamese's retrieval displays are often precipitated by human as well as by kitten behavior. Although she allows the Garcias to handle her kittens, she will retrieve them if the kittens' vocalizations indicate that they're being stressed. As the kittens mature and leave the "nest" on their own, the queen makes no attempt to establish a new home base. Many people-oriented queens give birth in their own special beds or even the owner's bed if that happens to be the cat's usual sleeping place. Unlike the wild queen, whose personal and physical territory are one and the same, these domestic cats may cling tenaciously to their personal sleeping quarters and refuse to abandon them regardless of what happens. More than one owner who delighted in the presence of an adult cat on the bed found sleeping in a feline maternity ward and nursery much less enjoyable!

Although most books on cat care usually state a specific period during which weaning occurs, in reality considerable variation may exist among different breeds or individuals. Beginning around the fifth week, the kittens rather than the queen initiate the nursing. From then on, two factors govern their feeding practices: the individual kitten's desire to nurse and the queen's willingness to provide this service. Ideally the queen's willingness to nurse complements the kittens' needs. In the wild any incompatibilities in this area would diminish the survival chances of the young, thereby eliminating the trait from the gene pool.

Fixed territories create fixed food reserves, so we can appreciate the delicate balance the wild queen must strike between keeping her young with her long enough to guarantee that they develop sufficiently, but not to the point of compromising her own food supply. This in turn suggests one reason why strongly people-oriented breeds and individuals may experience delayed weaning. Because people most likely served as the food source of those first protected Siamese mutations, there was no need for the queen to evict her kittens. She experienced no strain associated with five ever-growing young competing with her for food; the humans fed them all. And if the termination of weaning does depend on the queen's disappearing or driving the young out of her territory, those early Siamese who looked to their human "parent" for such a cue never received it.

As the wild queen provides her young with prey during the period of decreased nursing, a new form of feeding gradually replaces nursing, one that encompasses a completely different set of body-language displays. Instead of kneading and padding their food, the kittens now stalk and pounce on it, pinion it with unsheathed claws, and tear at it with their teeth instead of sucking. Simultaneously, as the kittens become more prey oriented and less milk oriented, subtle shifts in their personal space and social/asocial orientations begin to take place.

Compare this with what happens in the Garcia household, where the owners begin teasing the kittens to eat by pushing their faces into a bowl of soupy kitten food. Whereas the wild kitten has been primed by the queen over a period of weeks to recognize and accept food that provides tantalizing motion and sounds to peak its interest and appetite, the Garcia kittens must glean sufficient stimuli from an unmoving gruel in a stainless-steel bowl. Is it any wonder that Aquinas prefers to snuggle up to his familiarly scented, temperature-controlled, reassuringly purring queen-shaped food "dish" rather than compete with his larger littermates for such dull fare—especially since she makes no attempt to discourage him?

So, even though many experts cite six weeks as the average or even ideal age for weaning, it may normally occur much later in some breeds and individuals. Studies of some feral- or semiwild-cat populations indicate that weaning may not occur until eighteen months of age in some cases. While such durations seem extreme, cases where kittens left with their mothers continue nursing up to six months of age are not uncommon.

Suppose that a wild or feral queen gives birth to her kittens in early spring and continues nursing throughout the autumn and winter rather than weaning them in late summer. Because nursing inhibits the queen's heat cycle, doesn't this delayed weaning undermine that primary drive to perpetuate the species via reproduction? Yes; but in order for the species to survive, the *individual* must survive first. Under conditions of limited food supply and/or territory, delayed weaning provides two potent advantages that enable the individual to survive until conditions become more favorable. Not only does delayed weaning serve as an effective means of birth control, thereby limiting the number of animals competing for the limited food and/or territory available, it permits the most efficient use of that restricted environment. Compare an experienced queen hunting, consuming her prey, converting it to milk, and nursing her young, to an inexperienced kitten seeking scarce prey in that same limited and even hostile environment. Which animal uses its energy most efficiently? Which approach offers the kittens as a group the greatest chance of reaching physical (and reproductive) maturity? Via the passive process of delayed weaning, the cat effectively converts a hostile environment into one where both the individual and the species can survive.

Finally, delayed weaning maintains the social orientation, which decreases the probability of intraspecies fighting and gives the queen more time to teach her young to hunt under severe conditions or in a complex environment. The prolonged social orientation also favors survival where space is limited. Two queens, each nursing five eight-month-old "kittens," could share a small, isolated woodland or barn that twelve singular solitary cats would find intolerable. Kittens dwelling in harsh environments where prey is limited would learn the skills necessary to locate and capture their elusive prey over a span of months instead of weeks.

Given the combination of two factors—the survival value of flexibility in terms of when to wean plus the cat's tendency to respond to humans as queens—we can see how expecting domestic queens and kittens to adhere to a fixed weaning schedule borders on the ludicrous. Aquinas's mother looks to the Garcias as her food source (her "queen"), just as Aquinas looks to his mother for milk. Moreover, the Garcias *always* supply that food. Who can predict for certain whether the queen will respond like her ancient wild ancestors and refuse the kitten at a certain age or state of development? Perhaps she will be like her human "mothers," who willingly keep the food coming.

PARENTAL INDISCRETIONS

It's hard to believe that these relatively simple deep-seated natural mating and maternal displays could precipitate such a wide range of problem behaviors and human/feline misunderstandings, but they do. To appreciate the tremendous variety of sex-related problems fully, let's consider another example.

Jim Frankl and Toni Turelli, two graduate students, experience what they consider a most serendipitous event. A neighbor informs them that she just saw two abandoned kittens in the empty warehouse several blocks from their apartment building. "I bet the mother was hit by a car or abandoned them. Poor little dears barely have their eyes open."

Jim and Toni dash to the orphans, swiftly ensconce them in a flan-

THE BODY LANGUAGE AND EMOTION OF CATS

nel-lined box, and whisk them to the nearest veterinary clinic. Because Jim and Toni are pursuing advanced degrees in psychology and early child development, this seems like a heaven-sent opportunity for them to observe mammalian development firsthand. Although the veterinarian pronounces the kittens healthy, she urges the budding psychologists to return them to the warehouse and observe them from a safe distance for at least six to eight hours before assuming that they are abandoned or orphaned. "It's quite likely there were more kittens in the litter and the mother was in the process of moving them when your neighbor showed up," she tells Jim and Toni. She further stresses the crucial role that interaction with the queen and littermates plays in normal behavioral development.

However, Jim and Toni dismiss the vet's suggestion and warning, thinking them designed for people who lack their knowledge and expertise. They purchase some feline milk replacer and set themselves up as surrogate parents to the kittens they christen Oedipus and Electra. Unfortunately Jim and Toni's relationship hits a snag, and Jim moves out, taking Electra with him. In the following weeks Jim and Toni attempt to fill the void created by their ruptured relationship by concentrating all their efforts on their new pets. However, their approaches to rearing their charges differ markedly. Toni limits her interaction to the bare minimum during Oedipus's first months, because she believes that doing so will ensure that he retains his feline nature. Meanwhile, Jim becomes a doting father, responding to Electra's every sound and movement with slavish attention and concern.

By the time Oedipus and Electra celebrate their first birthday, their owners face a barrage of behavioral and medical problems. What few friends still visit Toni wear heavy boots and jeans. Most learned quite painfully that Oedipus will not tolerate any change in his environment and will attack without warning, sinking claws and teeth into whatever threatens him—including human arms and legs. Although he suffers numerous medical problems, trips to the veterinary clinic create such a horror show for cat and owner (to say nothing of veterinarian and

Feline Sexual and Mating Behavior: Too Close for Comfort

staff!) that Toni settles for home treatments and remedies. Even as routine a human/feline interaction as petting takes a bizarre twist. One minute Oedipus sits on Toni's lap allowing her to pet him at will; the next, he whips around, lashes out at her hand with his claws, then flees.

Across town Jim and Electra fare little better. Although not plagued by as many different medical problems, Electra never seems at ease. The least little disturbance sends her flying toward Jim, where she sometimes comes to a screeching halt and begins grooming herself vigorously. Because she grooms so much, hairballs cause nagging problems, and sometimes her excessive lapping produces unsightly bare spots and even oozing sores on her skin. Like Toni, Jim also sees fewer guests in his home because of his cat, but for different reasons. Electra becomes so frightened and upset by any intrusion that it takes Jim days to calm her down. He thought breeding her and letting her have a litter would mellow her, but the one and only liaison with another cat turned into a total disaster, so upsetting Electra that she licked herself raw and required medical treatment.

DECENT EXPOSURE

Can the problems plaguing Father Paul, Jim, and Toni possibly be related? To answer that question, we'll begin by listing the various behaviors displayed by the three cats:

- Sucking
- Dependency
- Hostility toward strange people or events
- Chronic medical problems
- Attacking when petted
- Excessive grooming and licking
- Fearfulness
- Intolerance of other felines

That's quite a behavioral rogue's gallery! However, if we evaluate these displays in terms of the cats' relationships to their owners and their

THE BODY LANGUAGE AND EMOTION OF CATS

environment rather than as isolated events, we discover that they can be perfectly normal under these circumstances.

"It is *not* normal for a cat to suck holes in clothing!" fumes Father Paul. "Nor to attack people for no reason at all," adds Toni defensively. "Nor to lick itself raw," chimes in Jim.

To be sure, we don't *want* our cats to behave this way. However, given these particular relationships in these particular environments, these particular cats are acting quite predictably and normally for them.

Must these owners resign themselves to living with seemingly neurotic pets? Not necessarily. But unless they understand both their relationships and the circumstances that led to them, they can't ever hope to alter the behaviors.

Before reading on, think about the list of negative behaviors and our discussion of normal mating and maternal displays. Recall the mating sequence, how the kittens nurse, and how the queen grooms and weans them. Then think about territoriality and the cat's sense of personal and physical space. Add to this your awareness of the incredibly powerful influence early experience exerts on feline behavior, and you gain some useful insights into why these animals might be acting the way they do.

THE SUCKING DILEMMA

One of my most painful memories from my days in private practice involves the frustration, anger, and helplessness I felt when the owners of a beautiful Siamese asked me to euthanize their pet because of his persistent sucking behavior. We tried everything—dietary changes including supplements and lanolin, punishment, companionship, and confinement. Whenever we thought the cat had abandoned the behavior, we inevitably discovered he'd simply found a more clever way to conceal it. In spite of all our efforts, I never felt that either the explanations for the behavior or the solutions recommended adequately addressed the problem.

Because these owners and their cat responded so patiently and cheerfully up until the end of this wretched and unfulfilling "treatment" process, I find myself still drawn to any article or information regarding this behavior. Unfortunately, many simply reiterate the same premises and solutions that proved so inadequate for me and my clients years ago. Then fate presented me with a sucking cat of my very own.

I was practicing in a college community at the time Eliot entered my life via a route not uncommon in such environments. Like many people embarking on a new life-style, college students often get pets for companionship, many of them "strays" that "just happen" to wander into the dorms. Because most college housing policies forbid pet ownership, sequestering the pet and evading direct responsibility for it pose the primary student challenges. Consequently we find more cats than dogs in dorms because the former are smaller, quieter, and thus more easily hidden. And because no one actually assumes ownership of the cat (and thus responsibility for violating campus policy), few of these animals are neutered and litters are common. Because queens often deliver their litters in the spring, these kittens become burdensome as the school year comes to a close, spurring the students to find homes for the kittens the instant they begin eating solid food.

Such was Eliot's history when he joined my household. Coal black and otherwise pathetically nondescript, Eliot possessed the most extraordinary set of feet. He belonged to that line of New England cats known for their polydactyly—extra toes—that make some of them look more like water striders then felines. In fact, Eliot soon earned the nickname "Thumbs."

By this time I had concluded that sucking occurred primarily in Siamese and offshoots of the Siamese gene pool. Consequently, it never dawned on me to look for this behavior in Eliot, who quickly filled out and assumed the solid, blocky conformation and easygoing personality of a typical domestic shorthair. However, a chance observation by a veterinary technician friend caused me to alter my views dramatically. My friend had marked the birth of my second son with a fringed woolen carriage blanket, which I draped over the back of the couch

THE BODY LANGUAGE AND EMOTION OF CATS

when not in use. When Eliot was about eight months old, my friend visited and asked, "What happened to the fringe on the blanket?" For the first time I noticed what was perfectly obvious to an outsider sitting on the couch: Much of the tapered two-inch fringe had been reduced to rounded nubs.

Even though I owned other cats, including a Himalayan, the timing and his age made Eliot the likely culprit. Pondering the situation, I remembered that my elder son often awakened in the morning with one cuff of his pajamas unaccountably soggy. I vowed to keep closer tabs on the now quite large Eliot. Sure enough, I discovered that as soon as Eliot thought my son was asleep, he would jump on the bed, seek out the sleeve cuff and begin sucking, kneading, and purring.

Unfortunately, two major characteristics of the classic sucking syndrome did not attend this behavior. First, Eliot obviously wasn't a purebred Siamese, and his relationship to that breed was remote at best. Second, he not only sucked wool, he sucked cotton, cotton blends, synthetics, or whatever fabric happened to encircle my son's wrist at bedtime.

At this point, I succumbed to anthropomorphism for possible clues to valid solutions to this behavior. What I saw was this big moose of a cat contentedly sucking the cuff of a toddler who was contentedly sucking a pacifier. Because I would never punish the behavior of my son, I couldn't justify punishing it in my cat. However, although I opted for acceptance, I continued observing Eliot as unobtrusively as possible in hopes of gaining further insight into the enigmatic behavior.

Early on, I concluded that the sucking evolved from a presleep pacifier to a stress-related behavior. Although Eliot outgrew the need to suck every day by the time he was a year old, if I caught him on the counter or strange people or events disturbed his daily routine, I'd discover him sucking that night. Toward the end of his sucking phase, when he was almost eighteen months old and weighed over fifteen pounds, he invented what we called "drive-in service." Rather than

Feline Sexual and Mating Behavior: Too Close for Comfort

settle down on the bed for the entire night, he'd stand on his hind legs and use his head to root around under the covers to locate my son's arm. Then he'd grab hold of the cuff with his teeth and drag it—and its attached arm—until it dangled over the edge of the bed. Then, resting on his haunches, he'd treat himself to a few reassuring sucks on the coveted cuff before heading outdoors to pursue normal nocturnal activities.

My experience with Eliot and the data on nursing and weaning in wild- and feral-feline populations now leads me to consider such sucking an extension of normal behavior rather than a physiological, nutritional, or psychological *problem*. Viewed in this light, the solution then centers around finding an appropriate way for the cat to fulfill this developmental need rather than stopping the behavior.

In my case, I simply let Eliot continue using pajama cuffs and blanket fringe; and generally I advise owners to allow the animal to continue using its accustomed article if possible. Given the stress-related nature of the behavior, removing the article or substituting another can often intensify and prolong the sucking. In the worst possible scenario, the owners make such a fuss about the behavior that the cat becomes a neurotic "closet" sucker, waiting until it's alone and then sucking slacks and sweaters with a vengeance. One owner quite literally discovered her sucking cat draped over the hanger that held her best suit!

What about the danger of digestive-tract obstruction from the ingested fabric? That a cat may suffer such a fate usually provides the major justification for smacking, squirting, spraying, or otherwise punishing or frightening sucking cats. However, doing so usually results in a self-fulfilling prophecy, because cats so treated tend to become even more frantic suckers who are more likely to shred and swallow fabric. Owners who take a more low-key and tolerant approach, however, seldom encounter the problem, because most of the wear and tear on the object results from kneading rather than sucking in nonstressful situations. Again, consider an anthropomorphic parallel: Compare contented infants and toddlers allowed to suck pacifiers or thumbs with those whose parents chastize and berate them for doing so. The former apply relatively little pressure and drift readily off to sleep; the latter appear literally to chew the object.

"I don't think the bishop would approve of my saying mass in a soggy robe full of holes," Father Paul notes wryly. Nor would other people whose cats zero in on heirloom afghans or expensive garments readily cede these to the cat without incurring negative feelings that could undermine the relationship. In such cases, try to find a substitute as close as possible to the preferred object. Apparently, sucking cats attempt to fulfill two primary criteria when they select a surrogate mother: texture and scent. Another familiar animal (preferably but not necessarily of the same species) might allow the cat to suckle. More than one owner has watched with fascination as his or her dog allowed a kitten or cat to knead and suck its abdomen. Other sucking cats will accept an animallike texture such as wool, particularly if it carries a scent the cat finds comforting. Depending on the particular cat, the appealing scent may be of animal or human origin.

In Eliot's case, the wool baby blanket initially offered acceptable surrogate scent and texture, but lost its appeal when it sat unused on the back of the couch during warmer weather. For whatever reasons—scent, convenience, identification with the pacifier-sucking toddler—he also accepted a variety of cuffs, but only those belonging to one specific individual.

I suggested that Father Paul ressurect his old sweater and present it to Aquinas with a minimum of fanfare, but cautioned him to secure all closets and leave no woolens lying about until he had ascertained his cat's willingness to confine his sucking to this one approved object. Because Aquinas slept with him, Father Paul taped a note to his bedroom door, reminding him to move the sweater from his den or living room to Aquinas's favorite spot at the foot of the bed.

If the cat sucks something too valuable to relinquish and you own no suitable substitutes, visit your local thrift shops. Often you can find

Feline Sexual and Mating Behavior: Too Close for Comfort

comparable used garments or blankets reasonably priced that will suit your cat's needs. (One client even discovered an afghan identical to the one her own cat was in the process of putting a hole in—only the one in the thrift shop had obviously already fallen prey to a sucking cat!) Wear the object or put it in the dirty-clothes hamper for a few days to absorb your scent. Remember, the cat makes an *emotional* as well as a physical association with the item and may very well shun something brand-new or smelling of moth balls.

What if you simply can't tolerate the behavior? Although those who advocate punishing the display might disagree, I recommend that disapproving owners find other homes for their cats, homes where the behavior will be accepted. Although some cats may abandon this behavior if consistently punished, others will manifest it in more severe or surreptitious ways. When this occurs, rather than seeking a new home for a feline "late bloomer," you wind up seeking a new home for a neurotic bundle of nerves.

One final note on sucking behavior. It's rare but I do encounter owners who allow their cats to suck on them. Although comprising a very small group, both cats and owners share some common characteristics. Almost all of these cats were bottle-fed by their owners, and almost all of the owners were single women. Because we know that neonatal kittens gravitate toward body heat rather than the taste or smell of milk, we can understand how some animals could easily begin sucking on fingers and other human body parts.

What are some of the drawbacks that can attend this kind of intimate physical surrogate-mother relationship? First, a person can never teach a kitten what another cat can. Our egos like to convince us we can, but a quick review of feline anatomy, physiology, and, above all, sensory ability and behavior tells us to forget it. Second, within the context of nursing, the queen also teaches the kitten to fend for itself; she doesn't promote infantile behaviors, because she doesn't want her young to stay babies forever. On the other hand, the people who get involved in intimate physical relationships with their cats often want exactly that;

they want the cat to be dependent on them and they encourage sucking as physical evidence of that dependency. Third and perhaps most troublesome, such relationships often become displacement behaviors for the owners as well as for the cats. The cat simply transfers its demands from the unavailable queen to the most readily available person. The owners then project their own needs onto the cat, needs that may be far more complex and emotionally charged than mere availability. But what happens when the lonely young woman who creates this intricate web of interaction with her pet suddenly meets a young man and falls in love? Now the cat's sucking on her ear or burrowing under the covers to lick more intimate areas becomes a source of embarrassment, guilt, and anger.

Obviously finding a new home for such a cat or curing its problems would require herculean efforts. A common rule we should apply to this and similar potential interactions is: If you can't perpetuate a behavior, don't initiate it. Anyone who's ever felt a newborn kitten latch onto a finger and begin blissfully "nursing" can surely appreciate what a powerful symbol of total trust and commitment the animal can become to a lonely human being. Those tempted to get caught up in such relationships can benefit from a more objective, unemotional understanding of the pitfalls of imbuing cats with qualities that actually lie beyond their ken.

Freeing Oedipus and Electra
Although orphan kittens raised by people without the benefit of other felines may display sucking behavior, other more severe behavioral problems usually occupy their owners' attention. Oedipus attacks and often bites anyone or anything he considers threatening. He's also plagued by upper-respiratory and urinary tract viruses. However, he responds so viciously to any kind of novel stimuli that taking him to the vet's for any kind of treatment generates an intolerable amount of stress in all concerned.

These behavioral abnormalities all result from the relative lack of

Feline Sexual and Mating Behavior: Too Close for Comfort

interaction Oedipus experienced with his own kind and his owner during those formative early months. Recall that Toni decided to interact minimally with Oedipus lest she hamper his "natural" development. Lacking all but the barest contact with others, he received no help in accepting his world. Left with no references but his own fears, he developed a fearful nature. Given the three fear options—fight, freeze, or flee—he opts to fight most of the time, hoping to "get them before they get me."

Although Jim took the opposite approach, becoming an almost constant doting presence in Electra's life, he also precipitates medical and behavioral problems in his cat, albeit of a different sort. Whereas Oedipus fights when frightened, Electra flees if possible and freezes if not. When stressed, she grooms, often licking and biting herself with sufficient force to remove hair and create oozing sores. Perhaps she's trying to recreate those earliest memories of peace and security when, as a newborn, her mother licked and groomed her; or maybe such displacement grooming represents an alternate manifestation of her otherwise unfulfilled sucking instinct. Nor can we overlook the possibility that lacking the experience or temperament to vent her anxiety on external objects, Electra attacks her own body to relieve the tension. One thing we do know, however, is that the behavior tends to be self-perpetuating. Because cats lick in response to the presence of fluids, and saliva present on the tongue tends to dampen the surface during licking, animals behaviorally primed to this response get caught in a vicious cycle.

While Electra bites and laps herself, Oedipus bites and chews on others—mostly people. Electra's medical problems—vomiting, diarrhea, hairballs, and skin lesions—rise primarily as a result of her displacement grooming; Oedipus's stressful worldview makes him a prime target for viral infections.

Despite their good intentions, Toni and Jim erred when they failed to recognize that normal feline behavioral development relies on a balance of two seemingly paradoxical extremes. Studies indicate that in laboratory and domestic animals, handling at an early age increases the individual's rate of development and ability to handle stress. In addition, other experiments indicate that lack of contact or isolation can completely disrupt an animal's ability to function normally even on the most primitive of levels. Yet other studies indicate that too much handling leads to stunted development, shyness, and a decreased ability to deal with novel events. These studies and others, rather than contradicting each other, indicate that normal development depends on a delicate balance of hands-on and hands-off interactions, a balance queens intuitively strike with their young, but one that often lies beyond our limited understanding.

Can Jim and Toni extricate themselves from the behavioral web entangling them and their pets? Because they initiated these abnormal patterns at such an early age and maintained them until maturity, both owners should undertake any change with the idea it may be a long process and the rewards may be minimal or even nonexistent. Above all, they should not attempt to rectify the feline companionship deficit in their pet's upbringing by introducing another cat into their household. Owners who take this quick-fix approach usually can't find enough negative adjectives to describe the results. In addition to all the antisocial behaviors already displayed by their cats, they must now deal with all those precipitated when one cat violates the territory of another.

In both Toni's and Jim's cases, they must carefully consider whether or not they can supply the highly controlled and limited environments necessary for their cats to feel secure for however long it takes them to develop more acceptable behavior. For the sake of the cats and the relationships, owners of such animals must honestly and objectively evaluate the four options _before_ initiating any programs. Accepting a cat who attacks your guests requires a lot more thought and self-confidence than accepting one who sneezes or passes gas; and the consequences could be much more serious. Attempting to change such behaviors could take months or even years. Terminating the relationship could plunge you into guilt; after all, if you had taken the time to find a foster mother with a litter, this never would have happened.

The problems, issues, and resolutions associated with raising orphan kittens are so numerous, complex, and emotional, that anyone who's dealt with them or attempted to help others deal with them can't help but recommend an ounce of prevention. I'd love to be able to say that humans make great surrogate queens, but that would be a bold-faced lie. The interaction between queen, kitten, and littermates reflects one of nature's most exquisite and intricate dances. Not only do cats march to the beat of different drummers, they respond to a natural music we humans can't even hear.

SLASH AND RUN

One of Oedipus's problems doesn't just affect orphan kittens, and it generates enough negative owner reaction to merit special mention. I call this problem display "slash and run" because of its unique form. Most commonly owners are calmly petting their cats when the animal suddenly wheels and lashes out with claws and even teeth. Although I can find little documentation on this behavior, my own experience combined with that of clients and colleagues suggests some possible explanations.

First, recall the mating sequence in which a male places his body across the female's back, then treads as he slides toward her vagina. After they mate and separate, she whips around, screams, and lashes out at her partner. Felines have followed this pattern for centuries, so it seems logical that cats instinctively know the sequence, especially since their asocial nature makes it unlikely that they learned the behavior by observing their elders.

When Toni rhythmically strokes Oedipus from head to tail, he looks deliriously happy, but is Toni inadvertently approximating the mating sequence, triggering her cat's intuitive awareness of an impending "attack" from the "female," causing him to lash out reflexively in self-defense? Similarly, female cats might also be instinctively primed to

The Body Language and Emotion of Cats

respond in the same manner to similar head-to-tail stroking. When they tire of such petting or when it exceeds a certain level, they terminate it with an inbred attack mechanism.

While this may appear unnecessarily vicious to us, we don't know that cats view this display as an attack at all. If they do, they certainly don't associate it with an attack by a predator, because no amount of postliaison violence will dampen their ardor or desire to seek each other, or another, out in order to mate again in the future. Although we can't prove that the female's swipe doesn't result from anger, we can't prove that it does either. However, one thing we do know for sure: If we happen to rub one of these sensitive felines too long or the wrong way, we can invite a nasty scratch or bite. Therefore, wise humans confine any petting to the head and ears of an unfamiliar cat until they learn to read its body language.

In the Eyes of the Beholder

To round out our discussion of mating and maternal-related behaviors and human responses, let's consider Robbie Bently, a mischievous preteen determined to liven up his mother's dull bridge club with the help of the family's three cats—Shadrach, Meshach, and Abednego. Robbie sprinkles catnip under the card table, then nudges the cats toward the women once they take their seats. At first the cats don't make a sound; they just roll euphorically under the table, occasionally wrapping themselves around a bare leg or rubbing against an ankle. The minister's wife is the first to wriggle uncomfortably; she bids way too high. Her partner giggles for no apparent reason, but Robbie can see that it's because she'd removed her sandals and Shadrach is blissfully licking her toes. Suddenly the third member of the foursome lets out a shriek and leaps out of her chair, upsetting the table and scattering the cards. "Something attacked me!" she screeches, clutching her bloody ankle. The three cats seem oblivious to the excitement as they roll and stretch, now emitting happy chirps and loud purrs. The minister's wife stares in open-mouthed horror at the spectacle; her partner giggles nervously to cover her embarrassment. The cardplayer with the bleeding ankle stomps out of the room in a fury. Carol Bently immediately recognizes the typical catnip response, which she normally finds amusing. She also recognizes that her guests don't share her opinion. Her look of confusion and consternation completes the quartet's emotional response, and Robbie can barely keep from laughing out loud and betraying his hiding place.

Whether we talk about nursing, sucking, or the various aspects of the mating display, the responses of Carol Bently's bridge club pretty well sum up the usual range of human reactions to these sex-related displays. One cardplayer reacts with shock and disgust, another with embarrassment, a third with anger. Carol considers her cats' behavior perfectly normal under the circumstances, while her son finds both human and feline responses absolutely hilarious.

Your own feelings may have run through the same gamut of emotions as you read various parts of this chapter. Nor did I escape some of these same emotions when I wrote it. "Will people be offended by this example?" I wondered. "Should I ignore the subject of intimate physical interactions between owner and cat rather than risk upsetting those who would never dream of becoming involved in such displays?"

To subordinate any doubts and present this material as objectively and realistically as possible, I reminded myself of all the times I've been on the receiving end of incomplete information when dealing with feline sex-related problems. On numerous occasions owners' emotions led them to hold back crucial information about the cat's history in order to disguise the exact nature of the problem for fear I'd think badly of them. Consequently, we wasted valuable time or, worse, implemented inappropriate solutions.

Ironically the more educated a person perceives himself or herself to be, the more likely that person will get enmeshed in problems of this sort. It was much more difficult for Father Paul and budding psychologists Jim and Toni to admit they had a problem and objectively evaluate all the contributing factors because in their minds they, if anyone, should be able to raise a normal cat. The fact that they failed naturally became a source of guilt and embarrassment. How well I remember the cheerful eight-year-old tugging on his mother's sleeve as I examined the family cat: "Aren't cha gonna tell her about Missy suckin' holes in everything?" he demanded. His mother turned beet red, swatted his behind, and shooed him out the door; then she apologized profusely for her son's "silly" outburst. Although I gave her several opportunities to discuss the problem and indicated my willingness to listen and help, she refused. I later learned from a mutual friend that my client feared that revelation of her cat's tragic secret would make me think badly of both her and her pet. So much emotion, and none of it contributing an iota to solving the cat's problem!

Whenever owners believe that their cat's behavior indicates a basic flaw, they often resent the animal for somehow trying to complicate their lives. This leads them to view the display as spiteful and deliberate, which in turn undermines their commitment to any solution. After all, who wants to devote a lot of time and energy to develop a good relationship with a spiteful cat that is deliberately trying to embarrass you?

BLIND LOVE

In addition to dealing with our own emotions regarding sex-related problems, we must often confront the opinions of others. Recall once again the delicate balance between the kitten's interaction with queen and littermates. When this balance goes awry, various developmental abnormalities result. We noted how the queen possessing poor maternal instincts probably fails to raise a litter successfully, thus preventing the perpetuation of that trait. Similarly, those who for behavioral or physiological reasons can't get the hang of mating obviously can't pass this characteristic on to future generations. In such ways, Mother Nature creates and maintains the healthy gene pool necessary for the survival of the best qualities of a species.

No one argues the usefulness of such a system for the wild feline.

However, we may not equate what's good for leopards, ocelots, and servals with what's best for Champion Bermuda's Triangle, the exotic shorthair. Instead, we may choose to convince ourselves that Triangle's extraordinary breeding requires special care and handling. If Triangle is psychologically or physiologically incapable of mating or taking a litter to term, or if she lacks the maternal instincts to raise healthy kittens, we may go to great lengths with controlled matings, caesarian deliveries, and hand-rearing to compensate for her feline deficits. How altruistic! We say, "Triangle is perfect in every other way, and it would be such a shame to deny her and the world her offspring."

Following this line of reasoning, many people feel that manipulation of the feline gene pool constitutes not only one of the primary joys of cat ownership but also one of an owner's primary *responsibilities*: "Exotic shorthairs provide all the marvelous qualities of the Persian without all the problems associated with that long coat." Others regard such human interference as the work of egomaniacs who wish to impose their beliefs regarding the "right" conformation, coloration, and temperament on another living being: "How could anyone create and promote a breed whose conformation results in chewing, eye and respiratory problems, and makes natural delivery difficult if not impossible?" (Before you exotic shorthair people send me scathing letters, please note that such extreme views have been and probably will continue to be applied to all breeds at one time or another. I use the exotic shorthair simply to illustrate my point.)

Such black-or-white judgments constitute a dangerously simplistic view. Both views contain elements of truth, but both deny *Felis domestica* its singular identity. Triangle is neither a wild creature nor an animate plaything for human amusement. She is a unique creature created by thousands of years of both natural and artificial evolution, and she functions best when her owners take into account all the complexities that attend her current position.

Theoretically, the idea of letting nature take its course makes a good deal of sense. However, those who espouse such beliefs more often than not actually expect the species to *revert* to its former conformation, coloration, and temperament. They fail to consider whether the genetic makeup that served the cat in the past best meets its needs now. For example, Aquinas cares little for other animals and would rather chase his ball than a bird or a mouse any day; but he absolutely adores people. In other words, genetically, physiologically, and temperamentally, he's the perfect parish cat. Once Father Paul realizes that these perfect qualities include the breed's tendency to form stronger and longer maternal attachments, which these cats readily transfer to humans and their belongings, he can readily accept and even celebrate rather than condemn certain disquieting variations.

Compare Aquinas to my cat Maggie, the typical brown tabby nocturnal predator. There is no doubt that if Father Paul and I left our pets to fend for themselves in the New Hampshire woods, Maggie's breeding would prove superior. But what about in Manhattan or San Francisco? Surely in such human-dominated environments Aquinas's bright-blue eyes, cheerful yowl of greeting, and willingness to approach people would lead him to a stable food source and reliable sleeping quarters much more rapidly than the aloof, human-wary Maggie.

So when we pass judgment on which cats should be bred and which shouldn't, which qualities we should perpetuate and which ones eliminate, we must always consider the animal's relationship to its actual environment and not simply adhere to some idealistic fantasy in which it runs wild or enjoys access to unlimited human intervention and veterinary care. Once we make a choice and pursue breeding programs to perpetuate certain qualities, we must also accept total responsibility for placing the resultant offspring in environments compatible with their physiological and behavioral makeup.

Above all, we don't play God; we put our relationships with our cats and our cats' relationships with their environments above our own egos. Before we congratulate ourselves on our ability to control feline genetics, anatomy, physiology, and psychology, we should remind our-

Feline Sexual and Mating Behavior: Too Close for Comfort

selves that the most influential Siamese mutation apparently just happened. We don't create; the most we can do is understand and then perhaps consciously or subconsciously manipulate what nature has already provided.

No less a scholar than William Butler Yeats noted that there are only two topics worth the consideration of the serious student—sex and death. The poet's comment arose not from a particularly prurient or macabre worldview, but rather from a keen awareness that these two aspects of life constitute nature's most fascinating and powerful forces. If we can understand and accept them as natural and nonthreatening, then we can understand and accept all of nature. Now that we know more about cats and sex, let's delve into that enigmatic specter of death that permeates the human/feline bond: predation.

8

PREDATION: THE CALL OF THE WILD, THE CRY OF THE TAME

*T*HE drama unfolds slowly one night in the crawlspace beneath the house. The mouse ventures forth from its burrow intent on reaching the break in the wall where it can scamper up to the kitchen. Maggie crouches in a corner about ten feet away, indistinguishable from the mottled black shadows cast across the uneven dirt floor scattered with old timbers and jutting rocks. Even if we could spy on both cat and mouse, we would see or hear almost nothing.

Maggie pounces so suddenly and effortless that humans must sense rather than actually see the motion, and within seconds the drama concludes as it has for centuries.

After what seems like weeks of rain, the sun breaks through, inviting us all outdoors. I sit in a lawn chair half-thinking, half-dreaming; the dogs spread themselves like shaggy throw rugs on the grass. Only Maggie seems interested in more active pursuits. I hear the chipmunk scream before I see it. Leaping from my chair, I fly to where Maggie holds the twitching creature in her mouth.

We've been through this before. She lets the chipmunk drop almost casually at my feet and glides away. My eyes fill with tears at the sight of the little creature's struggles. "Damn! Damn!" I curse both myself and my cat. Had I not interfered, the poor creature would be dead now, done in by that final well-placed bite. Now it rests in my hands,

Predation: The Call of the Wild, The Cry of the Tame

glassy-eyed and laboriously heaving in that predeath breathing the physiologists so precisely describe as "agonal."

Maggie sits on the stone wall carefully observing the scene like some ancient Egyptian goddess, her pupils almost imperceptible slits in the bright sunlight.

"Why don't you ever learn!" I shriek.

"Why don't you?" she seems to respond.

Nowhere in the consideration of feline behavior do the paradoxes assault us with such ferocity as when we study predation. Nowhere else do our negative emotions rush so powerfully forward to block objectivity and color our perceptions; nowhere else do we long to throw up our hands and admit, "I just don't understand!"

No one can deny the agony and sadness we experience when our cats down a fledgling robin or bluebird. Paradoxically, no one doubts the value of cats as rodent controllers without equal. From ancient Egypt to infested meadows in northern California, cats have proven their pest-control merits time and time again. According to some estimates, housing and feeding all the cats in animal shelters across the country and turning them loose to hunt would cost less than the amount spent to prevent and rectify rodent damage. Compared with most forms of rodent control, cats are prey-specific, nonpolluting, and nontoxic. Maggie keeps the house free of mice, yet I don't hesitate to hug her or let her sleep on the bed. Most rodenticides come with so many warnings on the package that I don't even want them in the house. Still, those poor chipmunks and birds . . .

Removing emotions from a discussion of predation is akin to asking someone to prepare a twelve-page recipe that begins, "Peel five pounds of grapes." However, peel we must if we're ever to enjoy the meal. We must understand predation objectively before we can fully enjoy the human/feline bond in its entirety.

THE PREDATORY DANCE

Separated from human emotions, feline predation's intricate and elegant body-language displays remind us of the Oriental dances of nature. When we divorce hunting and killing from any personal definitions of wrong or senseless brutality, we can see that the act itself consists of four phases:

· Stalking
· Catching
· Killing
· Eating

Every cat owner has witnessed the stalking phase, even if the cat in question has never actually brought down any prey. Picture that low-slung, rigid but flexible stance that enables the cat to move without seeming to move. Even people who despise predation must often admit that this instinctive display serves as a source of great entertainment and fascination during play sessions. Surely anyone with even the faintest interest in cats enjoys ever so slowly dragging a piece of string or ribbon across the floor and watching a cat pursue it.

The pounce and catch often occur so quickly that we see little more than a blur. Other times, when the cat appears to toy with its prey, this phase seems excruciatingly long. Although the cat's entire body moves with fluid coordination, front and rear legs perform quite different functions. With claws extended and tactile hairs erect, the front paws strike with deadly accuracy to pinion the prey. Meanwhile the rear feet remain firmly planted on the ground, providing a solid base on which the cat can pivot if necessary to block any attempted escapes or respond to changes in the prey's direction. This phase of the predation sequence also delights owners during play sessions: "Look at Muffy leap for her ball," Connie Farnsworth crows joyfully. "Isn't she the most graceful thing!"

The cat dispatches its prey, using its powerful canine teeth, or fangs. Most prey die almost instantly from a combination of shock, broken neck, and asphyxiation. If the cat intends to eat its prey, it usually begins almost immediately after effecting the kill. Unlike some wild felines, domestic cats don't normally store their prey or eat part of it and return to finish it later. We don't know whether this constitutes normal or circumstantial behavior in domesticated cats, because most pets have access to other food sources. The fact that a bowl of dry chow sits on the floor when they go out to hunt and will probably be there when they return undoubtedly influences the particular prey-eating displays we observe in domestic cats. This might explain why family pets who also hunt exhibit what their owners consider bizarre prey-eating habits. Maggie, the inveterate mouser, disdains mouse livers— or acts as if she does. On many occasions the only evidence of a successful hunt is the presence of a tiny rodent liver on the braided rug outside the kitchen door. The first time I saw one, my maternal instinct took precedence over disgust, and I automatically admonished her: "Mag, eat your liver. It's good for you."

Of course, it's also possible that these leftovers reflect not taste preference but remnants of the prey-gift display. Even if you don't own a predatory cat yourself, you probably know people who recount tales of their cats presenting them with gifts of dead mice, moles, voles, or worse at their feet or on their pillows. Such gift giving probably results from a cat's tendency to perceive humans as their mothers.

Recall the nursing sequence: Once the kittens' eyes and ears function and the animals begin moving around outside the nest, the queen brings dead prey to the area, sometimes consuming it in their presence. Then she encourages her kittens to eat the prey rather than nurse. Over a period of time she introduces stunned or weakened prey, which she corrals until the kittens figure out how to catch it.

Thus during the weeks of weaning, queen and prey become inextricably linked. Most significantly, the queen presents the litter with the dead game at the beginning of the sequence, when, we know, kittens most readily learn and imprint their behaviors. I wish I had observed Maggie's mother teaching her offspring to hunt, because I suspect that Maggie always ended up with the liver. If so, then unlike my sons, Maggie might not dislike liver at all. In fact, she may be saving the liver because she thinks that's what adult cats do. Now I, rather than Maggie, have become the beneficiary of that early eating lesson.

THE BODY LANGUAGE AND EMOTION OF CATS

Sometimes the prey remnants remind us more of trophies than of presents. It always amuses me to listen to some of my intrepid hunter-clients rave about "sneaky cats killing defenseless birds." I can't help but compare the mouse heads and flying-squirrel tails Maggie leaves in the wake of her own hunting expeditions to the stuffed deer heads and antlers in my clients' dens. Just as feline sexual behavior angers and repulses some people more for its similarities than for its differences from human conduct, those who berate the "sport"-hunting instincts in cats may be trying to stifle recognition of their own predatory natures.

THE PREDATORY RHYTHM

Like virtually everything that influences the human/feline bond, the actual steps of the predation sequence wield less power than their *relationship* to each other. When we explored the mysteries of the feline sensory system, we noticed that the amount of a stimulus necessary to trigger a system varies from species to species, and from system to system within a species. Cats require less scent, light, or sound to stimulate their olfactory, ocular, and auditory systems than humans. Among their senses, feline hearing exceeds the sense of smell in terms of sensitivity and functional utility. The more sensitive a system, the lower the threshold or the less stimulation required to trigger recognition and response. To get some idea of how thresholds vary among the senses, imagine your favorite Aunt Susan walking toward you on a bright sunny day wearing the same brand of lilac perfume she always wears. Because we humans rely more on our vision than our sense of smell, most of us could identify Susan by her image long before we caught a whiff of her identifying perfume.

Such thresholds also exist in behavioral displays. For example, a freckle-faced redhead I know will glower at you fiercely if you call him Red, and will probably come out swinging if you mention his freckles. As a parent, I can tolerate and even ignore my teenager's rock and roll up to a certain volume; beyond that threshold, I take action—"Turn that down!"

Predation: The Call of the Wild, The Cry of the Tame

Complex behaviors can oftentimes be divided into several steps, each with its own threshold, depending on the sense involved. As more and more stimulus occurs, additional facets of the response come into play.

Of the four steps in the predation sequence—stalk, catch, kill, and eat—stalking requires the least stimulation to prompt the response, and killing requires the most. What advantages does such a variable-threshold system confer? Although those who abhor predatory behavior might think of cats as near-perfect killing machines, in reality their success rate with mice (their prime target, toward which they display the *most* physical and behavioral sensitivity) comes in at a surprisingly low 33 percent. For every three mice the cat stalks, two will get away. And this represents the best the cat can do; when hunting grasshoppers, snakes, birds, houseflies, lizards, rabbits, chipmunks, and squirrels, the success rate drops even lower.

This fact immediately tells us that the cat who hunts to survive must spend more time stalking than eating, because at least two out of three attempts to secure a meal will end in failure. If the cat didn't initiate the hunting sequence until hungry, its relatively low energy level at the point would render it less able to perform the task at peak potential. For example, suppose you like fresh fish. If you know anything about fishing, you know that you don't put the pan on the fire, then saunter off to the riverbank with your pole to catch a trout while the butter melts. If you want fish for dinner, you allow yourself enough time based on your knowledge of the river and your quarry. You might even thaw a steak, just in case the fish aren't biting. And you would begin fishing well ahead of mealtime. If you wait to start until you're ravenous, chances are you'll get easily flustered and discouraged, maybe your hands will even shake. But if you allow yourself plenty of time, every cast can become a source of enjoyment. Similarly the cat goes stalking its prey long before it actually needs to eat.

Once the feline hunter locates a potential meal, what stimuli trigger it to pounce? When the prey stops moving, vision/motion data give way to hearing, which permits the precise localization necessary to direct a successful pounce. This explains why cats fail so miserably when hunting birds. Birds are "pecky" eaters, constantly hopping as they feed. Compare this fluttering to the stationary bulk-feeding rodents; once mice locate something edible, they munch away until they're satiated or the food runs out. No wonder many behaviorists and naturalists believe that the birds caught by cats must be ill, injured, or otherwise impaired; normal, healthy birds simply don't stand still long enough for the average cat to locate and attack them.

It takes more stimulus to trigger the pounce-and-pinioning response, just as it takes a harder tug on the line to get the fisherman to take notice. The fisherman drags his line through the water, and any tug less than a certain amount he ignores as inconsequential—a strong current, a snag of weeds perhaps. However, with a certain increase in pressure he immediately concentrates on his line to determine the right time and motion necessary to set the hook.

What kinds of additional stimuli increase the level to that required to trigger the kill response? As the cat pinions its prey, tactile hairs respond to any motion. The close proximity of the prey also generates useful scent data which further contribute to the overall stimulus load.

In cats the kill response differs markedly from an attack on another feline. When cats attack prey, no vocalization or threat display occurs; they don't hiss, they don't puff up their fur. They simply plunge their canines into the prey's neck. Compare this with the ritualized fighting between two cats where claws become the weapons of choice. First they square off, producing the identifying vocal and visual cues to inform each other of their presence and intentions. Next they swat with front paws, claws extended, aiming for the head and neck. If one of the combatants springs, the cats may lock front paws around each other, lashing out ferociously with their hind claws. Also bear in mind that intraspecies fighting depends on hormones to some degree; although females and all neutered animals may engage in such displays, we see them much more commonly among males. Predation, on the other hand, occurs with equal intensity among males and females, although

Predation: The Call of the Wild, The Cry of the Tame

queens with kittens tending to hunt the most proficiently.

Lack of awareness of this bit of feline combat style has led more than one person to concentrate on a fighting cat's mouth when he should have been paying much more attention to those rear claws. To be sure, cats may clamp with their teeth if sufficiently threatened or stimulated, but initially teeth function more as a holding than a biting device. Unfortunately when people suddenly realize that their fingers are in the cat's mouth, many panic and try to pull away. The increased pressure and motion merely supplies the additional stimulus necessary to trigger a more violent feline response.

In a typical scenario, the owner flips the cat on its back or gradually focuses petting in the abdominal area with the idea of "tickling Muffy's tummy." The exposed-abdomen position instinctively signals vulnerability to most mammals, and only the most stable cats will tolerate it for long. However, even these will often wrap their front paws around your arm when you commence the rubbing. Unknowledgeable owners will often see this display in a most positive light: "See how much Muffy loves this—she's hugging me!"

If the owner persists the hind paws come up, first with claws sheathed, then unsheathed if the person fails to take the hint. Next the fangs clamp onto accessible flesh or clothing and the claws begin raking it. At this point, *stop all motion completely* and remain as calm as possible. Screaming will only frighten the cat and perhaps switch your status from that of an antagonistic member of the same species to that of a life-threatening predator. If it's your cat, speak to it very softly, using its name frequently. Slowly bring your free hand up until you can gently scratch the cat behind the ears, then under the chin. Observe the pupils; continue gentle stroking and soft conversation until they look more normal (that is, not fully dilated or narrow slits). Don't try to rush the process. Remember, you "set" the cat with your own behavior and you must now allow it time to dissipate that stimulus if you wish to avoid a nasty scratch or bite.

Once the cat relaxes, gently open its mouth and extract your flesh or clothing if necessary. (Many cats automatically let go as soon as they relax.) Then carefully and gently extract the claws, all the while continuing your reassuring, soft patter. If you remain calm, the physical evidence of this encounter with a fighting cat may be no more than a few dents from the teeth and superficial skin pricks from the claws.

If you don't know the cat well, the most behaviorially sound but emotionally difficult response involves remaining as quiet as possible. Remember: Every sound or movement you make adds to the stimulus load and drives the cat toward a more serious response. Try to divert your thoughts elsewhere. Imagine yourself lounging on the beach or listening to your favorite music. Dissipate your own fear as much as possible. Most people who work with animals will tell you that dogs and cats sense fear in people and that this fear contributes to the level necessary to provoke antagonistic responses, just like any other stimulus.

Recalling the hierarchy of feline sensory development, it comes as no surprise that it takes the most stimulus to trigger a cat to eat its kill. Recall that taste functions simultaneously as a highly specific and yet a poorly developed sense; cats recognize "all right to eat" and "not all right to eat," but little else. However, as we noted earlier, considering the prey's habits and abilities, it would be a most inefficient system indeed that required cats to expend all that energy stalking, catching, and killing, only to discover that the prey flunked the taste test.

When we compare this phase of feline predation to our fisherman's response, we recognize that successfully landing the fish doesn't automatically mean it will wind up as tonight's dinner. Some fish may be tossed back because they're too small or inedible; others may be taken home for the cat (a gift) but not eaten by the fisherman. Others will fulfill all the legal, safety, and personal criteria and wind up on the table that evening. If these turn out not to meet the fisherman's expectations once prepared (that is, they flunk the taste test), the fisherman will not fish for them again. Finally, there will be those fish considered so special that they will get stuffed and mounted as trophies.

Predation: The Call of the Wild, The Cry of the Tame

A Curious Paradox

Studies of hunting behavior in many different species indicate that successful hunters display great curiosity. Thus, the cat's notorious curiosity seems perfectly reasonable. But, even this characteristic common to all hunters takes a paradoxical turn in the cat world. For if it is also true that curiosity kills, has nature bestowed cats with too much of a good thing?

Every veterinarian wishes he or she had a nickel for every scratched, bruised, or otherwise dinged-up cat whose owner attributes the mishap to the cat's curiosity; "She's always into things she shouldn't be. You know how cats are." Then whether we generally indulge in clichés or not, we find ourselves quoting the old saw, "Well, Ms. Farnsworth, they say curiosity killed the cat." Sometimes the owner will complete the adage: "Oh yes, Doctor, but satisfaction brought it back." We usually ignore this vital conclusion to the concept.

Whoever penned that elegant ditty knew a great deal about feline behavior. Cats are territorial; a territorial animal's survival depends on its ability to know its territory inside out, upside down, backward and forward. Every change warrants investigation. It might appear to be little more than an empty box or bag to Connie Farnsworth, but to Muffy it represents a perfect hiding place, not only for mice seeking to hide from her but also for her if she becomes prey. And of course, what a marvelous place to deliver kittens.

Studies of territorial awareness in mice show that animals who actively explore their territories increase their survival chances fivefold. Consequently, in addition to needing a strong sense of curiosity to protect it from being preyed upon by others, the cat needs this quality to track down an equally curious quarry.

So curiosity might, in fact, get our cats into a lot of trouble, particularly when they apply it to areas such as city traffic or the cleaning solutions under the sink against which they lack intuitive protection. Yet those who satisfy their curiosity not only learn and survive, they pass the trait on to their offspring. Ten years ago in my area we rarely heard the phrase "road savvy" applied to cats. We viewed traffic as alien to the feline way of life, and they violated it at their peril. However, as the human and pet populations and the traffic increased, owners began to notice that some pets learned about traffic more rapidly than others. The selective mechanism played itself out with brutal efficiency; those who claimed territories that included roadways but didn't learn about traffic became highway statistics. So, in a relatively short period of time, people began speaking of cats that either had or did not have road savvy, the same way they discussed their pets' mousing or maternal abilities. In such a way, the cat's natural tendency to explore every aspect of its territory led it to adapt to a completely alien "predator"—traffic. While we may argue the rightness or wrongness of this evolution, no one can doubt that a cat's curiosity enables it to adapt rapidly to environmental change.

BALANCING THE EVOLUTIONARY SCALES

I have already referred to the role territoriality and maternal displays play in predation. Another major influence contributes its share as well to the misunderstanding and negative emotions surrounding this behavior: We humans feel sorry for the cat's apparent victims. However, such feelings ignore a phenomenon some call reciprocal evolution. When two species interact as predator and prey, changes in one precipitate changes in the other. For example, I noted in the discussion of nocturnal behavior that the orientation to light or darkness of a cat's prey (or food source in the case of the house cat) determines whether a cat abides by diurnal, nocturnal, or dawn-dusk clocks. Similarly, the availability of food affects territoriality. When food is plentiful, cats may willingly accept less physical space than when food is scarce. This may explain why fewer cat fights erupt in crowded neighborhoods boasting large cat populations, except during the breeding season. Although a cat fending for itself in the woods probably couldn't survive in an area the size of a three-bedroom ranch, when that area contains

Predation: The Call of the Wild, The Cry of the Tame

reliably filled food and water dishes, it fulfills the cat's needs quite well.

Although we can certainly view reciprocal evolution as a never-ending attempt by the victim or prey to stay one step ahead of its adversary, we could also view it as a delicately balanced *pas de deux* within the ballet of nature. Viewed in this light, predator and prey work together for the beneficial evolution of both species into forms that best adapt them for survival in particular environments. Without such interplay, most changes would occur quite slowly, if at all, or as the result of random mutations. Compare the effects of someone tampering with the food in your refrigerator versus tearing up the road in front of your house: Which change would warrant the more urgent response? Suppose your do-list for the weekend includes mowing the lawn or repairing the hole in the fence separating you from your neighbor's vicious Saint Bernard: Which task will more likely stir you from your usual routine of watching the football game on television? When something as critical as food or personal safety is at risk, all but the most lethargic or inept respond. Those who don't suffer dire consequences, which in the animal kingdom usually include removal from the gene pool via death or failure to reproduce.

No one can doubt the efficiency of such a system; and certainly a biologist can appreciate its beauty. Ever since Isaiah referred to the lion lying down with the lamb, however, people have perceived predation in a negative light and longed to institute a peaceable kingdom wherein all animals treat each other as best buddies. We simplistically *choose* to perceive predation as evil and wrong; we look at unemotional reciprocally evolving predator/prey pairs and decide that this occurs *only* because the predator can't tap alternate food sources. If we could provide that alternative, we tell ourselves, we could stop the wretched display.

However, we can't deny that if we disrupt the predator/prey interaction, we also destroy the accompanying incentive that results in change. Without the presence of its evolutionary partner, predator and prey may both suffer. Pause a moment and consider the degree of mental and physical coordination that accompanies the typical predator/prey interaction. Is it any wonder that animals kept in captivity or otherwise fed artificially and deprived of this stimulus soon become dull and lethargic? Does it really surprise us that they succumb more readily to disease and physiological ailments?

Or consider what happens in my own home. When we first open up the crawl space under the house every spring, it appears that Maggie will surely kill every rodent living there. However, closer examination reveals that the first casualties are the extremes: the very young, the old, or the unfit. The healthy, smarter ones survive and multiply, and after that first spree Maggie's success rate drops precipitously, then gradually increases as she refines her hunting skills to match those of her more wary opponent. Compare this response with that of bottle-fed Ray, who never caught or ever even met a mouse. One night I heard a commotion in the kitchen and discovered a very confused, frightened Ray lying on the floor, his front paws wrapped protectively around his food dish. Less than a foot away, a terrified mouse ran in circles squeaking loudly. Evidently Ray had surprised the mouse accidentally on his way to his food dish, and the mouse instinctively responded with fear. Lacking any predatory training and evidently few mouse-attuned predatory instincts, Ray responded in a most uncatlike manner. All he cared about was protecting his food; he perceived the mouse as a predator after his crunchies rather than as prey for himself. Meanwhile, it appeared that Ray's lack of response also aborted the normal rodent display, causing the poor creature to become locked in its frantic race to nowhere.

Humans who disrupt the predator/prey interaction often fail to recognize that it serves a very useful and necessary purpose and that its removal creates a void in the natural process of change. We also overlook the fact that nature inevitably fills such voids, often using people to fill the vacuum created by the removal of prey from the natural equation. More often than not, people replace the prey as the stimulus

Predation: The Call of the Wild, The Cry of the Tame

for change. Instead of evolving strong limbs or a thicker coat in response to a faster outdoor rodent prey, the cat evolves a shorter nose or silkier fur to fulfill the human concept of beauty or novelty. The reason we often deny this realization arises from the obvious dilemma it creates: In this new equation, who is the predator and who is the prey? Who is the victim and who is the villain?

Feline evolution has been linked to the predator/prey relationship for centuries; therefore, it would be foolish to suppose that the need for this balance or an outlet for these displays would instantly vanish just because we remove one of the components. More often than not, the surviving component will fulfill the needs previously met some other way—and some of those alternate routes may be even less acceptable to humans than others. Like other owners of hunting cats, I dread when I have to keep Maggie indoors. Because she's accustomed to all the physical and mental activity that accompanies the hunt, she finds my small house quite boring and turns to me for attention. She stalks me as I move through the house; she pounces on laundry as I try to fold it; she swats paper and pen as I try to write. In short, she makes a general nuisance of herself compared with her usual low-key, well-mannered behavior. On the other hand, I know more than one house-cat owner who considers what I refer to as Maggie's nuisance behaviors among their pets' most endearing displays.

ONE STEP AT A TIME

Once we understand predation as a precise four-step process in which the two participants respond to each other to perpetuate or terminate the display, we can see the tremendous variety inherent in what we once considered a rigid (and negative) behavior. Moreover, we can see how transference of any part of the display from prey or predator to human can result in an equally wide range of behaviors and attendant human responses.

We can now also appreciate why cats appear to kill "for the fun of it." When we consider the behavior unemotionally, we realize that the

THE BODY LANGUAGE AND EMOTION OF CATS

cat simply hasn't received sufficient stimulation from the activity to carry it through the eating phase. What are the survival benefits of this? If you dropped your fishing line into the water and a large trout immediately bit the hook, shook it weakly once or twice, and then went belly-up, how anxious would you be to eat that fish, even after you had cooked it thoroughly? Perhaps the "thrill" of the hunt or chase also serves as an indicator of the physical fitness of the prey. Prey that gives in too easily may be suffering from conditions that might prove harmful to the predator.

One kind of prey that appears to contradict this hypothesis is birds. We have already mentioned that the majority of the birds cats catch are most likely debilitated in some way. How can the cat derive sufficient stimuli from such weak prey to trigger consumption and why doesn't the cat get sick? Because birds are normally much more active than rodents, a sick bird and a healthy rodent could conceivably supply the same amount of stimulus. And because birds belong to a completely different part of the animal kingdom than cats or mice, and possess their own unique physiology, conditions harmful to birds— diseases, poisons, parasites—may pose little or no threat to the cat. In addition, most birds are diurnal, so cats free to exercise their limited-light and nocturnal preference in response to the habits of the rodent population wouldn't normally come into contact with them. Those cats that do adhere to the diurnal life-styles of their owners may give the erroneous impression that they *only* hunt birds and that they *always* kill and eat what they catch.

The stepwise nature of predation also explains why well-fed cats may finish a meal and immediately go outside and begin stalking. It takes so much more stimulus to trigger eating than it does stalking that the latter may strike the cat as a completely separate act. Unfortunately, some owners simply can't see it that way. Every time they see the cat displaying the stalking behavior, they assume it's hell-bent on killing and eating something, and they yell at it or chase it with a hose. This reaction always reminds me of unconfident wives married to men who

Predation: The Call of the Wild, The Cry of the Tame

admire beautiful women. Although their husbands never do anything but look, these wives insist that admiration must *always* lead to an illicit romance. Such lack of faith and nagging can actually cause the unwanted behavior, and I wonder if the same holds true for cats. Could the owner's excitement at such times actually contribute to the amount of stimulus necessary to turn a casual stalk into a kill?

Looking at this phenomenon in a more "positive" light, many house cats routinely dive for their food bowls after a play session. Although they may need physiologically to replenish the energy expended during the activity, the behavioral reason may be that the typical stalk, catch, and pounce games create enough stimulation to trigger the eating response.

Let's return to the crawl space under my house and apply what we know about the need for ever-increasing stimulation to carry the predation sequence through to completion. As soon as Maggie creeps into the area, she conducts a sensory sweep to begin the stimulus-collection process. At this level a slight movement or sound will trigger a change in her posture. She freezes and uses her ears to localize the source of the movement. If it doesn't recur, she continues her exploratory prowling; if the stimulus repeats itself, she assumes the much more deliberate stalking posture that conceals her own presence while allowing her to continue collecting maximum information about the environment.

If she receives sufficient data to pinpoint the mouse, she moves in its direction. Obviously, if the mouse continues making sounds or motions, these will intensify as Maggie closes the distance between her and her prey. At each point during this process the failure of the mouse to provide a stimulus could abort the sequence. However, as Maggie inches closer and closer to her prey, the chances of this happening decrease proportionately.

For example, if Maggie thinks she hears something else ten feet away and freezes so that she'll hear the sound if it's repeated, but the mouse remains perfectly still, the cat will probably give up and resume her more generalized prowling. On the other hand, if she receives a similar sound cue from only a few inches away and both she and the mouse freeze, there's a much greater chance she'll now pick up other sensory data—such as odors or tactile sensations—that were beyond her capacity when she was ten feet away.

Although sufficient levels of stimuli can turn cats into precise killing machines, insufficient stimuli can make them look inept and even stupid. Once, I observed Maggie practically sitting right on top of a chipmunk, totally oblivious to it even though she seemed to be staring right at it. Mother Nature and protective rodent coloration and behavior combined to save the chipmunk's life. Initially drawing the cat close as a result of some indiscreet sound, the plucky chipmunk froze like a rock. Simultaneously a light breeze came up, further directing any sound or scent away from the cat and creating a lot of distracting foliage motion. Although to me—and probably the chipmunk—Maggie's gaze seemed fixed right on it, she eventually moved off, and the chipmunk scurried away.

STACKING THE DECKS

Feline predatory behavior may interest humans, but human reactions to it can be downright fascinating. For every owner who admires a cat's hunting skill, there's another who prizes a cat because it never hunts. For every owner who begs, "How can I teach Muffy to catch mice?" there's one who pleads, "Tell me how to make Muffy *stop* hunting!" Death always strikes a nerve.

When asked for my advice on guiding cats toward or away from predatory displays, I quote that old song that urges us to "accentuate the positive, eliminate the negative, and don't mess with Mr. In-Between." If we want a cat that doesn't hunt, we must take the time and effort necessary to track down a breed and individuals known for their lack of interest in this activity. If we remove the kitten from the queen as early as possible, thus ensuring that minimal hunting skills will be transmitted from parent to offspring, we stack the deck in our favor. If we want a hunting cat, we should apply all our effort to selecting a kitten born of known hunters and allow it to remain with the queen long enough to permit maximum transfer of hunting skills and knowledge. That stacks the decks the other way.

Simple enough, but several factors conspire to thwart these approaches. For one thing, we know that separating queen and kitten early to avoid transference of predation-related information may also result in a pet with other behavioral deficiencies that may prove more troublesome than hunting. Similarly, leaving a kitten with its mother long enough to guarantee its hunting abilities may produce a highly asocial, territorial, and nocturnal as well as predatory animal.

Finally, but most critically, many of the body-language displays we find most appealing and even encourage in our cats stem from predatory behavior. Look at your cat's toy collection and the array available at your supermarket or pet store; peruse the ads for feline paraphernalia in your favorite cat magazines. So many of these items capitalize on kitty's desire to stalk, chase, and pounce—and kitty's owners' desire to see such displays—or else they provide a convenient perch, covered hideaway, or "cave" from which to initiate such activities.

When Connie Farnsworth dangles the fuzzy toy mouse on a string just out of Muffy's reach, little does she realize that she's encouraging her pet's predatory behavior. All Connie (and most of us) sees are those endearing movements, those lightning-fast leaps, those successful (or unsuccessful, if we pull the prey/toy away fast enough) catches. In other words, much of what we call "playing with the cat" consists of our providing a predatory stage for our pet's natural responses.

There's absolutely nothing wrong with such play, provided we understand our role in nourishing or minimizing the predatory instinct. Most studies indicate that even when raised with absolutely no external source of predation-related input such as that given by the queen, 50 percent of all cats will automatically initiate the hunt sequence when presented with a mouse. Furthermore, nearly all the nonhunters will join in the hunt when placed with a predatory female.

While this would seem to provide conclusive proof of the depth and strength of the hunting instinct, we must remember that these findings only have meaning if the researcher's interaction with these test kittens didn't stimulate or reinforce the predatory behavior in any way. Because the human inclination to urge a kitten to stalk and pounce runs so high, experimenters could eliminate this variable only by raising the animals in total isolation. But because such isolation produces such a wide range of pronounced behavioral anomalies, the conclusions of such studies would offer little Connie could apply to her relationship with Muffy.

Whether the origin of the predatory display lies solely within the cat or depends on the presence of prey or prey-substitute play, it does constitute a major portion of what we call normal cat behavior. When problems arise, more often than not they involve a change in environment or human rules regarding acceptable forms the display may take. For example, Connie and Muffy thoroughly enjoyed their chase-and-pounce games in their Chicago apartment, but when Connie moves to a large condominium complex complete with fenced yards, the previously agreeable human/feline relationship begins to unravel.

For one thing, Connie feels she moved to the condominium *for* Muffy because she believed her active and inquisitive pet needed more room to explore. She introduces Muffy to the "great outdoors" of her protected yard and patio with a great sense of pride and accomplishment. After checking out every corner of the enclosure, Muffy settles herself beside Connie on the patio; contentment floods her owner as she sips her drink and basks in the late-afternoon sunlight. Connie completely misses what Muffy noticed immediately on her rounds: The previous owners of the condominium had routinely left seed out for the birds. Not only have the birds become used to feeding less than ten feet from the brick patio, they believe the area to be totally safe.

For Muffy, it amounts to little more than the feline version of shooting ducks in a barrel. She snares and kills the finch so quickly, it most likely had no idea what hit it. Connie experiences a similarly dramatic response when her pleasant, self-satisfied reverie is shattered by that one wretched, suddenly aborted squawk. She opens her eyes and spies her beloved, delicate and dainty Muffy, feathers and blood soiling her otherwise immaculate white bib.

A cat first perceived by an owner as a loving, benign, nonviolent, playful companion that unexpectedly turns into a vicious killer inevitably strains the human/feline bond. Humans may respond with the full range of negative emotions: anger, revulsion, frustration, fear, guilt, and even hatred. Then, depending on our individual personalities and temperament, we convert these emotions into some sort of body language expression. Connie leaps from her chair, shrieks, and heaves her thick paperback novel in Muffy's direction. I, too, leap at the sound of captured prey, but I know that yelling at Maggie and charging toward her serve no useful purpose. Consequently, I try to take a more introverted approach; my stomach churns, my whole body goes rigid. Regardless of how owners manifest these emotions, they do little to strengthen our relationships with our pets.

When the heavy book slams into her rear end and Connie's shrill cries pierce the air, Muffy immediately responds defensively, fleeing for the safe haven offered by the stack of empty boxes and packing crates by the door. At this point owners typically respond one of two ways. The fact that the cat has not only killed a defenseless creature but has also run away may intensify some owners' negative feelings. In this situation, they seek out the cat with the idea of meting out verbal and/or physical punishment. Other owners may feel guilty about their own response and seek out the cat to console it.

If Connie rips into the pile of boxes with the idea of teaching Muffy a lesson, she may or may not succeed in achieving her goal. But whether she does or does not, her relationship with Muffy will suffer. Suppose she succeeds in flushing Muffy out of her hiding place after five minutes of rummaging, swearing, huffing, and puffing. Grabbing the terrified animal by the scruff of the neck, she shakes her roughly and screams, "Don't you ever touch a bird again!" Then she swats Muffy's behind and drops the cat unceremoniously to the ground. Given all we now know about the feline senses and behavior, can Muffy possibly connect all this unnatural confusion to the expression of a perfectly *natural* instinct minutes ago? Hardly. More likely, the cat will associate the hubbub with Connie rather than with the bird. So, rather than discouraging the predatory behavior, the human response simply undermines the relationship.

What if Connie fails in her attempt to find and punish Muffy for her transgression? Muffy might be able to evade her owner long enough for Connie to expend her negative emotions some other way, perhaps calling her best friend and pouring out her feelings over the phone. Or in her fright, Muffy might leap from the stack of boxes to the top of the stockade fence and disappear into the next yard. Or she might become sufficiently threatened by Connie's behavior that she attacks her owner when cornered.

If Connie feels guilty about her response to the incident and seeks out her pet to console her, she creates another unproductive situation. Essentially Connie now winds up responding positively to a display she personally finds reprehensible. As she sits by the pile of boxes, cooing softly to her cat, stroking her in all her favorite places and promising her a bowl of ice cream, will Muffy connect this behavior to the fact that Connie doesn't want her to hunt? You know the answer to that.

MAKING THE BEST OF IT

Experience and conversations with cat owners lead me to conclude that the *worst* time to initiate an antipredation campaign is when you catch your cat red-pawed with a dead bird or mouse. Although classical behavior-modification theory might propose this as the best time to dispense any punishment, several factors work against this approach to predation. First, the owner would essentially be punishing the cat for something it considers normal behavior. (Imagine our fisherman quietly savoring the taste of his trout and his vegetarian neighbor rushing in and belting him with her purse: "Take that, you wretched killer!")

Second, human emotions flicker from cold to hot so rapidly that few people can exert enough self-control when predation occurs to guarantee a *consistent* response. (One day the vegetarian calls our fisherman a killer, the next she tells him he's ruining his health, the third she accuses him of deliberately trying to make her unhappy, and the fourth she decides to try a more "positive" approach and invites him to her house for dinner.) Third, we know that cats don't respond well to punishment. They freeze, fight, or flee, signaling their fear of the individual meting out the punishment; however, they may not link the punishment with their own behavior—especially if that behavior springs from deep-seated instincts. (If our fisherman cherishes his mealtimes alone, he may resent his neighbor's intrusion to the point that he doesn't even hear her message.)

Given the dearth of productive responses, I prefer acceptance. So when I see Maggie with that limp body in her mouth, I try not to leap from my chair, but that doesn't mean that my heart doesn't leap to my throat. By now I recognize all of my body-language expressions; I feel the tension and the anger, and one by one I consciously try to dissipate their adverse effects. It wasn't easy at first, and some days it's easier than others.

Initially and still on some occasions, I calm myself by recalling Albert Schweitzer's espousal of "reverence for life." Those with only a passing acquaintance with this philosophy often erroneously envision Schweitzer as a man who found any death save by old age intolerable. Not so. Among his many talents, Schweitzer was a physician, and he routinely used whatever drugs or treatments he needed to destroy the microorganisms or parasites threatening the lives of his patients. He argued not for the preservation of every living creature, but rather that no death go unnoticed.

The teaching of Native American cultures and others more attuned to living with nature than most technological societies, with their attitude of humans-as-outside-observer-and-judge, can also provide great comfort as we seek to accept predatory behavior at the moment of the kill. Reciprocal evolution is a contemporary concept, yet many "primitive" cultures have recognized it for eons. Every true hunter knows that we are all predators and we are all prey in the overall scheme of things. Therefore it's virtually impossible to pass judgment on the predator or prey status of another individual or species without passing judgment on ourselves. To ignore or participate in this process unconsciously is not to think; and without thought, there can be no reverence for life.

A Pound of Prevention, an Ounce of Cure

Although prevention presents the best cure for most problems, it doesn't necessarily work wonders when applied to altering hunting behaviors. We already noted that even if we did everything "right" to eliminate all predation-related learning from a kitten's upbringing, the odds that it will resort to the behavior when exposed to suitable prey are still fifty-fifty. Second, such a restrictive upbringing might very well create even more troublesome problems. And finally, eliminating human/feline interactions which include predation-related displays would eliminate much of the unique feline body language we so enjoy. If Connie could extinguish Muffy's stalking, pouncing, leaping, tossing, and pivoting displays, would the result still fulfill her definition of a normal cat?

The best cure for predatory behavior demands total confinement in a prey-free environment. Obviously you can more easily achieve this state in a brand-new, climate-controlled apartment on the twentieth floor than in a loosely constructed summer home in the woods of New Hampshire. However, most owners who choose to confine their cats wind up consciously or subconsciously using the confines of house or apartment to divide the cat's world into "fair" and "unfair" game. Bugs and spiders are usually considered fair game, but birds or squirrels are not. Stalking the former may even earn the cat the owner's applause or laughter; stalking the latter wins it a scolding or spanking.

Theoretically we could build a case for not indulging the cat in any

Predation: The Call of the Wild, The Cry of the Tame

chase or pounce games or giving it any mouse- or prey-shaped toys. Even if practical, this has little to offer because the resulting animal wouldn't be what most of us consider a cat at all. It would act more like a dull overtrained dog or, worse, like a robot. In either case the animal's behavior would be so aberrant, we would have to question the purpose, ethics, and humanity of such an approach.

Finally, we can certainly work toward extinguishing the behavior via selective breeding. By mating poor or disinterested hunters we can noticeably improve our chances of producing cats with similar non-hunting traits. When we opt to manipulate the feline gene pool, however, we must exercise great caution. If Muffy proves to be a notoriously poor hunter—"She couldn't catch a mouse if it bit her!"—we owe it to ourselves to ask why. Why is Muffy so unnaturally inept at performing one of the strongest innate feline displays? If we conclude that "she prefers her ball," then we must proceed knowing we're merely transferring the target or prey preference to something that coincides more with our own definitions of acceptability. If we want Muffy and her progeny to play ball in the house rather than hunt in the woods, we must also guarantee an environment where the cat can continually manifest that behavior: We mustn't breed cats content to play ball in apartments and then sell them to people who want a good mouser to live in their barn.

Suppose that Muffy's poor hunting ability results from her poor eyesight, weak hindquarters, or insufficient mental ability to capture prey? Selective breeding could perpetuate such undesirable, possibly detrimental traits to achieve the desired behavioral results. However, breeding physically or mentally deficient animals simply eliminates the behavior by default. Unlike the cat who stalks, chases, and attacks a ball rather than a mouse, progeny of ill-conceived breeding programs based on the perpetuation of flaws that decrease the probability of the cat *successfully* displaying the behavior often don't lack the urge or desire to do so. To my mind, such approaches rank with extracting all of a cat's teeth to keep it from biting. They are even worse, because at

THE BODY LANGUAGE AND EMOTION OF CATS

least the toothless cat's frustration and confusion dies with it: Those rendered ineffectual hunters by virtue of poor coordination, eyesight, or respiratory problems perpetuated by selective breeding pass these weaknesses on to countless generations.

THE OTHER SIDE OF THE COIN

Suppose instead of moving into a spotless condo, Connie and Muffy move into a hundred-and-fifty-year-old Victorian that's been unoccupied for five years. As soon as Connie sees the house, she knows that a little paint, wallpaper, and landscaping will turn it into her castle. However, during the absence of human occupants, mice, squirrels, and bats have laid claim to various nooks and crannies of the structure. The first night Connie leaves a loaf of bread on the kitchen table, she awakens to find only crumbs remaining.

"Muffy, hurry up and get settled. You have work to do," Connie semijokes, waving the empty bread wrapper at the fluffy white cat carefully grooming herself on the old brick hearth. "There are tons of mice waiting for you." Muffy pauses midlick to stare quizzically at her owner, her deep-blue eyes taking on a faint turquoise gleam in the sunlight.

In the four months that follow, the mice destroy ten rolls of specially ordered wallpaper and chew the mats of two limited-edition prints. Connie loses her patience: "You stupid cat, you better start earning your keep around here! It's the least you could do after I let you lounge around the apartment all those years."

Trying to make a hunter out of a nonhunter takes as much skill as attempting the reverse. More often than not owners once again wind up struggling against genetics, the environment, or both. If the cat lacks the interest or the physical or mental ability to hunt as a result of accidental or deliberate genetic manipulation, striving to turn it into a hunter promises the same results as those achieved when trying to change the proverbial tiger's stripes.

For example, suppose Muffy's unique appearance results from the

Predation: The Call of the Wild, The Cry of the Tame

intermingling of some very people-oriented Persians and some highly inbred blue-eyed white domestic shorthairs. Although she appears perfectly normal to Connie in every way, Muffy's genetic makeup may compromise her ability to hunt in several ways:

- Her vision may not be optimally developed for limited-light activity.
- Her hearing may be limited in the ranges where her prey makes most of its sounds.
- She may be more vocal, thereby alerting prey to her presence.
- Her white coat provides zero camouflage.
- Her long fur impedes motion.
- Her people orientation makes her prefer diurnal rather than nocturnal activity.

From this list we can see that the very genetic qualities that made Muffy such a perfect pet in the downtown Chicago apartment render her a wretched country huntress. Connie could never hope to beef up Muffy's physical and behavioral attributes to successful predator levels. Even if appropriate technology existed, it's doubtful Muffy would accept corrective lenses, a hearing aid, and a mackerel tabby sweater, to say nothing of psychoanalysis geared to help her abandon her preference for daylight human companionship in favor of nocturnal rodent hunting. Given these conditions, Connie would be much better off investing in a few mousetraps.

A different cat might make a better mousetrap, provided it came from proven hunting stock. This would seem the ideal solution, because the hunter will often teach the nonhunter any lessons the latter didn't get from its mother. Those who favor this approach recommend that the nonhunter be placed with a queen and her kittens if possible, because these females make the best teachers, if for no other reason than that the queen must hunt more often to feed her young.

If I appear to be wording the second-cat solution carefully and even a bit evasively, that's because this approach flies in the face of all we know about territoriality. Introducing a strange cat to a nursing queen could precipitate more fights than teaching sessions, but amazingly this seldom appears to be the case. It could be that in addition to lacking an awareness of predatory body language, nonhunting cats also lack knowledge of those postures used to challenge members of their own species. If so, the queen may relate to Muffy as a kitten and vice versa. If Muffy's reluctance to hunt arises from her lack of knowledge and experience rather than innate physiological and/or behavioral limitations, we can expect her to return home and assume her rodent-control duties.

However, if Muffy is strongly dependent, devoted, and owner-oriented, she may not appreciate being shunted off to a strange home with strange cats. Once there, she may hide under the bed and refuse to eat or drink, and once back home she may seek to reaffirm her territory by spraying urine or clawing Connie's new wallpaper. Equally negative results may occur when a strange cat comes into the dependent cat's household for the purpose of rodent control.

The predatory display is strongly entrenched in wild felines because immediate survival depends on the animal's ability to catch prey. However, as domestication makes hunting ability less critical, other behavioral displays may become more so. Cats whose food takes the form of little stars deposited in a bowl on a mat in the kitchen may feel much more protective about that kitchen than the cat who catches mice throughout a fifteen-room farmhouse and the fields surrounding it. Because of these often subtle differences in orientation, you should approach introducing a new cat into your household or sending your cat to another household to learn hunting skills with an open mind. Ideally, nonpredatory felines will welcome another cat into the household to perform those duties they personally find unacceptable for whatever reasons. Ideally, those nonhunters placed with other cats in other households for educational purposes will learn to hunt, or at least will not suffer unduly from the experience. But we can never be certain what will happen until we try. Only a basic awareness of each cat's individual personality and body-language expression can help ensure the desired results.

Predation: The Call of the Wild, The Cry of the Tame

GETTING THE BIG PICTURE

When I took driver's education years ago, the instructor constantly hammered into us the need to "Get the big picture. Don't confine your field of vision to the straight and narrow, seeing only your side of the road and that area immediately in front of and behind your car." This concept holds even more promise for those of us trying to deal with our feelings about predatory behavior in our cats.

Whenever we judge predation as wrong, wicked, or abnormal, we trap ourselves and our cats in a relationship equivalent to sitting in bumper-to-bumper traffic with an infinite double yellow line stretching out ahead. Such beliefs and their corresponding emotions and body language create such a limited range of normal and acceptable behavior for our cats that only ignorance can sustain it. But as soon as we recognize that much of the body language that so delights us when we play with our cats springs directly from their inherent hunting instincts, we begin reevaluating our entire attitude toward the display. Considering the number of people who totally and emotionally condemn predatory behavior, it's obviously easier to choose emotional ignorance over unemotional and objective knowledge. However, the resulting deeper relationship and stronger bond make it more than worth the effort.

In addition to recognizing the basic physiological and behavioral components of predation, getting the big picture also means recognizing how this display relates to all the others we've studied. A cat with a strong hunting instinct will also instinctively coordinate its life-style to that of its prey. Thwarting that preference will not eliminate the predatory urge but merely shift its focus. Owners who believe they can make their hunting cats nonhunters by confining them at night may discover that their mouse hunters want them to play at 3:00 A.M., or prey on chipmunks and birds by day.

In terms of mating and maternal displays and their relationship to the hunting instinct, our paradoxical feline doesn't disappoint us. We would expect those potent male hormones that induce toms to go at each other so enthusiastically to influence their hunting ability, yet such

THE BODY LANGUAGE AND EMOTION OF CATS

does not appear to be the case. On the other hand, those nursing queens so gently attending their young prove themselves to be the most skilled and enthusiastic hunters within the feline population.

When we further expand our big picture to include early kitten experiences, we can see how nature reinforces a relatively specific predator-prey relationship. During that limited time when kittens cement their food preferences, the queen will obviously present them with the most readily available forms of prey the most often. The kittens, in turn, will experience the most education in the location, capture, and consumption of that particular prey. Thus, the queen doesn't train her young to be indiscriminate killers who attack everything that moves. Rather, she teaches her young the four-step sequence as it pertains to the available food supply. Compare this with the owner who laughs and cheers every time his or her six-week-old kitten attacks a piece of string or ribbon, toy mouse, or fringe of a bath towel, then cries when he or she gets the veterinary bill for emergency surgery to remove the cellophane wrapper or tinsel from the cat's stomach: "How could a cat be so dumb?" Similarly, when the cat launches itself at a three-year-old girl's white socks—and the ankles within them—with all the enthusiasm it displays toward the white balls of crumpled paper or fabric tossed or dragged provocatively by its owner in play, unknowledgeable humans accuse the cat of being vicious. Whereas the queen teaches her young to hunt in harmony with their environment, unknowledgeable owners often create indiscriminate killers and call it play.

When we frame our big picture with an awareness of territoriality, we must once again admire the rigid-yet-flexible balance that exists between the two displays. The hunting and territorial instincts are closely intertwined, but they manifest differently from individual to individual. Poor and good hunters alike may become highly territorial. While those who rely strongly upon their human companions for food may feel extremely protective of their owners and their owners' immediate surroundings, those who depend upon a certain area to fulfill their predatory needs may feel equally protective of that physical space.

Consequently, although a knowledge of species behavior can lead us to understand these critical body-language displays, how they relate to each other depends on our intimate knowledge of our own particular cats. To recognize one but not the other would be like driving a car looking *either* ahead *or* in a rearview mirror, but not both.

When we stand back and examine the big picture created by the interrelationship of predation with the other physiological and behavioral facets of the species and each individual cat, we can't help but be overwhelmed by the most exquisite paradox of them all: The ever-present specter inherent in predatory nature breathes life and uniqueness into our relationship with our cats. Maggie's graceful leaps and pounces remind me daily that death shares the lead in this perfectly choreographed and balanced dance we call life.

Having wound your way through the body-language displays of cats and our human responses to them, you may have discovered that some of your personal belief about cats in general and your own cat in particular were reinforced or challenged. Those reinforced should increase your confidence in your relationship with your cat and enable you to enjoy it even more. Those challenged should make you think. Rather than being indicative of our inadequacies or failures, the challenges we confront and resolve often exert the strongest influence on our bonds with our cats and provide the greatest understanding and joy. Many times that understanding comes from science, but mythology provides that unique form of joy. In the next chapter the paradoxical feline invites us to create a view of individual ownership that enables us to unite the best both science and mythology have to offer.

9

FELINE SCIENCE AND MYTHOLOGY: DEALING WITH CRISES

*I*T was a sad day when Copper died. Dear lovely Copper, age seven, not at all old for a cat, had succumbed to a lifetime of obesity and all the wear and tear fifteen extra pounds of weight had visited on his vital organs over the years.

During those seven years Copper had become a fixture in our town, his name, picture, or actual presence an integral part of everyday life. The town paper heralded the beginning of the Christmas season with a portrait of everybody's favorite fat cat wearing a Santa Claus suit, his huge head wreathed in snowy whiskers. His Easter photo portrayed everyone's feline pal as Super Bunny, complete with long floppy ears, happily munching a chocolate egg. At Halloween, he didn't even need a costume; the rotund orange cat looked like the Great Pumpkin with no makeup at all.

By the time I met Copper, he was already a legend, an enormous twenty-nine-pound local celebrity. More knowledgeable and experienced practitioners than I had discussed the value of weight reduction with Copper's owners, Dick and Dottie Harcourt, but all to no avail. We all failed because no amount of scientific data and predictions could extricate Copper from his mystique as the town's adorable fatty. Even to his owners, Copper's expansive, lovable personality and expanding girth were inextricably related; to decrease his weight would be to diminish his personality. Because they and a surprising number of

Feline Science and Mythology: Dealing with Crises

others made this "jolly fat cat" association, Copper grew bigger and bigger. Then one day the giant orange cat refused to eat.

His owners immediately rushed Copper to the veterinary clinic. Although under normal circumstances going off-feed for a day hardly signals a feline health crisis, in Copper's case it not only signaled a crisis, this time it signaled the end.

Even though such clashes of paradoxical and extreme beliefs constitute a routine aspect of most human/feline interactions, the memory of Copper's death, his owners' heartache, and the sympathetic outpouring from the community that loved that big cat left an indelible mark on me. Pondering how human mythologies come to support what I must call "abuse" or mistreatment of the most subtle, insidious kind fueled my desire to write this book; and even now during those quiet hours when Maggie curls up beside me, my thoughts drift back to Copper.

What went wrong? The Harcourts certainly weren't stupid or ignorant people. Nobody doubted that they loved Copper and wanted only the best for him. But he did die at an age when most owners are just beginning to appreciate the full range of their relationships with their cats, and he did die from what can only be called man-made complications. How did this happen?

Although specific body-language displays and strictly behavioral problems can strain our relationships with our cats, feline physical ailments can wreak the most havoc in our lives. Ironically, even though the idea of Mousey succumbing to leukemia or needing an abscess lanced or a fractured leg repaired puts us through an emotional wringer, we rarely give credence to the role our emotions play, not only when our cats get sick but also in maintaining their health.

VITAL SIGNS

As novice veterinary students, my classmates and I were required to learn the necessary procedure for conducting a proper physical examination and making an accurate diagnosis. I can still recall one professor, fire in his eyes, demanding, *"Tell me what's wrong with this cat."*

THE BODY LANGUAGE AND EMOTION OF CATS

A confident student volunteered, "He's too thin." Another added, "His coat's dull and dry." A third piped up, "His mucous membranes are pale." Emboldened by the professor's deadpan expression, the rest of us rushed to contribute our own negative appraisals. Before long, we had attributed every problematic or potentially problematic physiological and/or behavioral symptom in the book to this poor creature. We could hardly wait to begin saving its life! We watched our teacher's face, fully expecting his praise and admiration; instead he paled, then slowly turned the color of a ripe apple. We barely had time to prepare for the explosion.

"How can you know what's wrong with this cat when you don't even *know* this cat? How do you know he's not a *normally skinny cat with a dull, dry coat and pale membranes!*" The man seemed to increase in size and intensity, not unlike an angry cat himself. "You can never know *abnormal* unless you know *normal* first!" His words cut like a laser through all our preconceived, often nebulous notions and etched a new principle in all our minds.

Try this little experiment. Without looking at your cat, answer the following questions:

- What color are you cat's eyes and coat?
- Describe the quality of its coat and skin, its teeth and gums.
- What sex is the cat? Neutered or intact?
- How much does it weigh?
- How much does it eat and drink every day?
- When, how often, and how much does it urinate and defecate?

Did you know all the answers? Like most owners, you probably don't know at least one basic fact about your cat. Scan the ads for lost animals in the newspaper and you'll discover a lot of other loving owners who can describe little more than the bare minimum about their pets. I used to live in a housing development where free-roaming cats outnumbered the kids, and the kids were legion. On an average day I'd see at least four black-and-white cats performing normal feline activities throughout the neighborhood. Consequently, when a handmade

Feline Science and Mythology: Dealing with Crises

poster bemoaning the loss of a black-and-white cat showed up on the phone pole at the end of my street, I couldn't help but picture those owners greeting a stream of kids, each returning the "lost" cat. As soon as the owner would tell the child that the cat he or she found on Laura Lane wasn't the right cat, the child would release the animal and it would scamper off to Maple Avenue, where another child would "find" it and truck it to the owners.

Once we think about it, the major dilemma confronting us when we face this problem becomes obvious: How can you tell that a cat is lost? Only the most dependent house cats *look* lost, and these seldom stray very far from their sheltered environments. Because few lost cats display the body-language signals (crying, begging on strange doorsteps, seeking out unknown people) that we humans associate with being lost, only a detailed physical description of the cat will help others identify it.

The inability of owners to describe their pets adequately also explains why knowledgeable shelter workers demand that those who have lost pets visit the facility to look for themselves. Attempting to match an owner's over-the-phone description with one of the animals in the shelter not only takes time, it seldom produces a perfect match. One person's mackerel tabby may easily be another's brown stripe; one's purebred Persian another's "angora."

And even when owners do come in to identify their pets, the system isn't foolproof. My favorite example of the description dilemma involves a client who had two great loves in his life: beer and his cat. One day he weaved into the waiting room carefully carrying a long-haired calico cat. "This my cat?" he slurred politely. No, it wasn't; his cat was a mackerel tabby shorthair male. He returned this "lost" cat to the shelter, where eventually, via trial and error, he located his own pet. When owner and pet were finally reunited, the owner sincerely rejoiced: "Am I ever glad that's over. I really missed the little guy!"

Even when we know certain facts about our cats, we often find it hard to articulate them:

THE BODY LANGUAGE AND EMOTION OF CATS

"I can picture exactly how Tippy's coat feels, but I just can't describe it."

"I know I'd recognize Sonora immediately even in a roomful of cats, but I just can't put it into words."

"I can't tell you what's 'normal' for Sassy, but I know immediately when something's bothering her."

Not recognizing or being able to describe specific feline physical characteristics or body-language signals causes no problems as long as the cat remains within the owner's definition of normal—however nebulous and ill defined that definition may be. But when things go wrong, this lack of clarity may jeopardize our chances of solving even the simplest problem.

For example, in the months preceding Copper's demise, his owners brought him in for a complete physical several times because "He's just not right" or "He's off his feed." Although a perfectly legitimate observation in their minds, it presented me with pitifully little to work with; in my mind, this grossly overweight cat hadn't been physiologically "right" for years. While the Harcourts saw the problem as a single abnormality undermining their otherwise perfectly normal cat's health, I struggled to figure out which of the abnormalities precipitated by his excessive weight was the most likely culprit this time.

The Harcourts believed that I could magically diagnose Copper's "problem" using state-of-the-art blood tests, X-rays, electrocardiograms, or whatever other technological marvels I might have up my professional sleeve. They simply didn't realize that in order to maintain Copper's "fat cat" mystique over the years, they had consistently chosen to accept certain abnormalities as normal for him.

Therefore, when Dottie so earnestly asked, "Could Copper be coming down with pneumonia? His breathing seems labored at times," she reflects her belief that the problem arises from some outside assault on her otherwise perfectly healthy pet—some virus or bacteria. But I knew that Copper's heart had been straining to supply blood and oxygen to all those extra pounds and that this, in turn, had led to a certain amount

Feline Science and Mythology: Dealing with Crises

of fluid accumulation in his lungs. Believe it or not, the Harcourts knew this too. They saw the X rays; they listened attentively to my explanation and cautions about weight control; they even medicated Copper to relieve the symptoms. However, in spite of their awareness of Copper's basic physiological deficits irrefutably traceable to his weight, they consistently refused to acknowledge this connection.

"That's horrible (mean, stupid, selfish)!" you may say, but we all adjust our definitions of normal, no matter how slightly, to suit our own beliefs about our cats. Maggie has an undershot jaw—her lower jaw protrudes more than her upper one—which gives her a distinctive appearance I find not only normal but also appealing. Others might not share my view of this structural variation. A professional breeder might euthanize such an animal the instant the flaw became apparent. Another person might feel negatively about the jaw but positively about the cat's other qualities and seek out a veterinary orthodontist to correct the problem with braces or surgery.

Similarly, some owners don't mind if their cats play host to a flea or three, whereas others can't tolerate the presence of any parasites. Some of us see a bit of earwax as a normal feline secretion; others will take a morning off from work to take the cat to the vet to have its ears examined.

In short, what we perceive as normal or healthy in our cats consists of an aggregate of factual and personal data. Therefore when something goes wrong, both owners and veterinarians can find themselves entrapped in what often turns out to be a most paradoxical approach to physical and behavioral health. The Harcourts bask in the public acclaim heaped upon them and Copper due to their cat's enormous size, but they don't like how that excess weight adversely affects his health. Jane Doe considers fleas a normal part of cat anatomy, but she finds the presence of the tapeworms transmitted by those fleas repulsive beyond words. Her husband, John, considers mousing a normal feline behavior; however, he smacks the cat anytime it deposits dead mice on the doorstep.

THE BODY LANGUAGE AND EMOTION OF CATS

WHAT'S NORMAL?

Back in the turbulent 1960s and 1970s those of us in college at the time often whiled away the hours pondering the imponderable: What's normal? What's real? Years later I thought I found the answers when I read *The Velveteen Rabbit* to my sons. In this beautifully written children's tale, I learned that to be real is to be loved. Then, during a discussion about canine behavioral problems with a most knowledge-able trainer some time later, I discovered that when trying to establish a stable bond between owner and dog, love is *not* enough.

Initially this paradox horrified me. I had always maintained that love conquers all, sustaining owners through even the worst medical and behavioral crises. However, once I recovered from the shock of this assault on my romantic notions, I heard what my friend was *really* saying. He wasn't saying that the love between owner and pet isn't important or necessary; he was saying that when things go wrong, we need something more: unemotional knowledge. And when can we best collect facts unemotionally? When things are going right.

Let's consider three owner descriptions of their perfectly normal cats. To Dick and Dottie Harcourt, Copper's "normal" includes his twenty-nine-pound weight and a coat that they must brush daily and bathe weekly to keep glossy and odor-free. Copper moves slowly, never climbs stairs or jumps up on furniture. During periods of temperature or humidity extremes, he has trouble breathing. Copper always re-sponds to the sound of the can opener or refrigerator door and rarely misses a meal. His normal diet includes anything the Harcourts eat, in addition to his own feline fare. Periodically his stomach gurgles; he passes gas and may vomit or have diarrhea.

Maggie normally maintains her weight in the optimum predator range; you can't "pinch an inch" on her. I keep her bowl of dry food full and have an undefinable but quite distinct idea in my mind how much she eats. During the summer Maggie hosts a few fleas, and her coat becomes somewhat dry and brittle in one small area at the base of her tail. She usually sports a few dings and scratches, but her ears and teeth are spotlessly clean, her eyes bright and free of any discharges. Sometimes she sneezes once upon entering the house; and she occasionally burps once after eating.

Paul Weatherby owns a blue Persian, Adonis, whose pedigree makes many a cat fancier green with envy. Paul weighs Adonis biweekly, immediately increasing or decreasing the amount of the specially formulated diet he feeds his pet if the Persian's weight varies more than an ounce. Play sessions with Adonis occur twice daily and consist of a specific set of activities designed to exercise all major muscle groups and develop coordination, as well as challenge the cat mentally. Paul's concept of the normal Adonis includes the Persian's enthusiastic response to these sessions and his ability to participate in all the activities for the recommended amount of time. Paul grooms Adonis daily and bathes him every six weeks. During these interactions he compares Adonis's eyes, ears, claws, skin, and mouth to his mental ideal. It's normal for Adonis's gums to be pale pink; it's normal for his eyes to tear. Strong odors almost invariably make Adonis wheeze and sneeze, and he frequently vomits small amounts of foam or semidigested food and hair.

As you read these descriptions, did some of these "normals" strike you as odd? Some probably did. However, although we can pick out the oddities in others' definitions, we often fail to recognize similar oddities in our own. So how do we sort out the myth from the science?

Within limits, your veterinarian can supply the most objective data about your cat's normal anatomy and physiology. I say "within limits" because in spite of all our marvelous technology, much of medicine still deals with ranges and probabilities rather than absolute values. The "average" cat may have a body temperature of 101.5 degrees F, but your cat's temperature may normally hover around 102.8.

While the limitations of the equipment or technique may account for certain variations, we must also recall our discussion of normal biological cycles: It's more normal for Adonis's temperature to vary over a twenty-four-hour period than not. If we don't make the same measurements at the same times, any differences must be *obviously* different to be significant. If Paul establishes 102.2 degrees as a normal temperature for Adonis by taking the Persian's temperature every day for two weeks when he grooms the cat at 8:00 A.M., this doesn't automatically mean that the cat is running a fever at 4:00 P.M., when his temperature registers 102.6 degrees. However, while Paul's veterinarian may not worry much about that 4 P.M. reading, he or she would undoubtedly consider a reading of 103.6 degrees significant at *any* time and seek to determine possible causes for this abnormal elevation.

Similarly the "normals" for active predatory felines differ from those of sedentary house cats. Nursing queens exhibit normal physiology and behavior unlike that of spayed females or those in heat. That which is defined as normal kitten nutritional requirements or bone structure would signal serious medical problems in a twelve-year-old cat.

A second factor that may limit your veterinarian's ability to predict normal for your specific cat relates to his or her opportunity to see your pet under normal conditions. No, you don't have to invite Dr. Blackbag to your home for the day so that he can observe Spunky eating, drinking, urinating, defecating, and playing with her toys. On the other hand, it's infinitely easier for practitioners to eliminate irrelevant variables when diagnosing sick animals if they've previously seen them in healthy states for routine vaccinations and examinations.

For example, Maggie hates to ride in cars, despises her carrier, and accepts strangers only on her own terms in her own sweet time. Were I to take her to even the most skilled practitioner in the most perfectly appointed clinic, normal for her would include an elevated temperature, rapid heart rate and respiration, and dilated pupils. Other cats may urinate or defecate under stress; still others may drool, vomit, have diarrhea, or shed copious quantities of hair.

Unless the veterinarian and owner have experienced the physical and/or behavioral changes that occur when the normal, healthy cat visits the veterinary clinic, they could misinterpret these as signs of a medical problem. If Paul stuffs an unwilling Adonis into a carrier and rushes him to me because his temperature has risen 0.2 degrees F and the terrified cat upchucks his dinner on the way, a frustrating sequence gets set into motion.

Paul explodes into the clinic, "Quick get a doctor! Adonis has some terrible disease. He's got a fever and vomited all the way over here." As part of the examination, I take Adonis's temperature. "It's a hundred and three," I report. "Oh my God," shrieks Paul. "It's gone up almost a degree in less than an hour. Hurry, Doctor, do something quick!"

We can certainly appreciate Paul's anxiety, but try walking in my veterinary shoes for a minute. My training gears me to beware of jumping to conclusions regarding the abnormal unless a valid aware-ness of normal exists. Add to this other training that compels me to conduct certain procedures or tests and dispense any medication *for a specific reason*. That means that I conduct tests and dispense medication only if the animal's condition warrants it.

If I've seen Adonis for routine vaccinations since he was a kitten and know that he always arrives at the clinic in a frantic state, I may concentrate my efforts on calming Paul down once I've ascertained that the cat's symptoms most likely arose as a direct result of the stressful trip. Ideally, I'll be able to meld my own objective observations with Paul's emotional ones to obtain the most complete picture of the cat's condition. If careful questioning reveals that the only abnormality prior to the upsetting trip was Adonis's 0.2-degree rise in temperature, I may send the cat home with special instructions for Paul: Observe food and water intake, urination, defecation, and note any sneezing, cough-ing, vomiting, or diarrhea, and so on. If I feel that Paul's emotional state would render him unable to make such determinations objectively and without upsetting the cat or himself, I may suggest that Adonis remain at the clinic for observation.

However, suppose the first time I see Adonis is when Paul bursts into the clinic with the wide-eyed, vomit-covered, feverish cat in tow. Now I must rely on those far-from-perfect textbook "normals" and Paul's ability to describe what's normal for his cat. If the cat has never ridden in a car or visited a veterinary hospital before, even the most observant and indulgent owner can do little more than guess at how many of the signs are normal reactions to the car ride and how many are related to any medical problem.

This creates a catch-22 for the clinician. On the one hand, I certainly don't want to perform unnecessary tests or dispense unneeded medications; not only would such serve no medical purpose, they could cost the owner quite a bit of money. On the other hand, I don't want to send home a sick cat that actually needs diagnostic tests and medication; while this approach doesn't cost the owner as much money, it could cost the cat its life.

Seen in this light, understanding feline body-language expressions and the role our emotions play suddenly takes on very practical and serious dimensions. Unless we know what's normal for our cats and can extract our emotions from these evaluations, we may hamper even the most caring and sophisticated attempts to solve any problems.

Begin now, when things are going well, to get a strong sense of what's normal for you and your cat. Read any of the many books on basic cat care and contact your veterinarian, breeder, or groomer to fill in any blanks. Then when problems arise, you can concentrate your efforts on maintaining a solid bond rather than sorting through conflicting facts and negative emotions.

OVER AND OVER AGAIN

Two kinds of problems exert the greatest pressure on the human/feline bond: those that are chronic and those that are sudden, traumatic, and life-threatening. Common chronic problems include:

- Urinary tract infections
- Food allergies
- Teeth and gum problems
- Skin problems
- Matting and grooming problems, hairballs
- Upper-respiratory problems

Feline Science and Mythology: Dealing with Crises

Although cats seldom die from these conditions, more than one human/feline relationship has suffered when they persisted day in and day out, year after year. Moreover, chronic problems always possess the potential to escalate into critical ones at any time. All owners of male cats prone to cystitis know that every episode can turn into a life-threatening blockage regardless of the steps taken to prevent this.

Consider Paul and Adonis: With frustrating regularity Adonis comes down with cystitis whenever Paul kennels him or takes him to strange places. Fortunately Adonis hasn't blocked since his first bout with the virus three years ago. However, that incident scared the daylights out of Paul and prompted him to switch his cat to a special (and expensive) diet, to change Adonis's litter daily, and to medicate him at the first sign of any urinary tract problem.

Although we can say that Adonis is a basically healthy cat, we must also admit that Paul spends a great deal of time worrying about his cat's health. Because his work requires that he travel a fair amount, it seems he's forever worrying that Adonis will get sick. At times as he carefully measures out Adonis's food and medication, he envys his neighbor's relationship with her ragged shorthair, a cat that eats everything and has no consistency in its life whatsoever—but has never been sick a day in its life.

Similarly those whose cats develop allergies to specific foods must often wrestle with feelings of resentment, frustration, and guilt. Suddenly the dream of the idyllic relationship with Puff evaporates with the realization that you're going to be feeding her a specific brand of chicken-based canned cat food for the rest of her life. When you see a photo of the golden-haired toddler sharing an ice cream cone with a fluffy kitten, you feel cheated; three laps of ice cream would give Puff diarrhea for a week.

In addition to those gnawing physiological problems, owners must often confront those resulting from negligence. Because I find the only product that successfully controls fleas on my pets—a powder—personally distasteful, I always postpone flea control until it's too late. Then,

The Body Language and Emotion of Cats

in addition to dealing with the despised powdering process, I must also contend with the pets' scratching and any problems related to that scratching (hair loss, skin irritations), not to mention the army of fleas hopping around the house. Further complicating matters, I must also deal with my awareness that I, more than anyone, should know better.

I can tell you from personal experience that the cat who offers resistance to the owner attempting to rectify his or her own negligent behavior need do very little to precipitate an emotional crisis. Like many owners who powder their cats, I find it easiest to powder a towel and then rub Maggie with it. However, if she dives under the couch when she sees me coming or tries to turn it into a game, gleefully leaping from couch to chair to shelf while I mark her passing with clouds of powder on the furniture, my patience quickly ebbs. And God forbid I lunge and entrap her only to have her rake my arm with her claws in the process: Then, in spite of all my objective knowledge, I'll burst into tears.

Foolish? Of course; I can be foolish and so can any loving owner who must treat chronic problems in their cats. Those negative feelings and how we manifest them may defy the best logic; but that doesn't make them any less real or potent. The only way to set aside our schizophrenic emotions and limit their negative effects on our bonds with our cats is consciously to accept full responsibility for the treatment of the chronic problem and all our feelings about it. And in order to do this, once again we need more than love; we also need knowledge.

Taking Control

It still amazes me how many owners address the diagnosis and treatment of any feline problem as though it lay totally outside their control. Although such a "Dr. Smith made me do it" approach may sustain most owners for the ten days necessary to get rid of Sassy's ear mites, it doesn't offer much support to the owner whose diabetic cat requires constant monitoring and medication. Even if Dr. Engelbert embodies the most exalted, qualified, angelic, or fearsome veterinarian on this earth, eventually going to all that (emotional as well as physical) effort *for him* wears thin. Unless we can say with certainty, "I'm doing this because it's the right thing for Sassy and me and because *I want to do it,*" we will surely undermine the positive emotional bond in the process of treating the problem. If Paul keeps Adonis on a special diet "because the vet insists," but chafes every time he must drive out of his way to buy the food, Adonis may grow physically healthier, but his relationship with Paul may deteriorate.

Whenever you confront a physical or behavioral problem, be sure to find out what the treatment involves *before* you begin. Ask questions, explore options and alternatives, seek a second opinion if necessary. Don't worry about upsetting others or looking stupid: Qualified professionals prefer knowledgeable and committed clients. If you find this other person difficult to communicate with for whatever reasons, that can only make things worse. It doesn't matter if Dr. Jacobson is the best in the business; if you personally find her arrogant and rude or feel that she treats Puff too roughly, either change your feelings or find another veterinarian. If your own experience with cortisone left you with some very strong negative feelings about this drug and Puff's veterinarian prescribes it for her allergic flare-ups, confront the issue now. Such soul-searching should not be viewed as complicating your life, but rather simplifying it: to put you in control of what could be a long-term situation right from the beginning.

THE STRESSFUL PARADOX

When Adonis succumbs to his first urinary tract infection and subsequently blocks, Paul experiences a wave of emotions, all of them negative. He feels negligent because an unusually busy schedule allowed little time for much interaction with Adonis. He feels guilty because he missed the early warning signs and he's frightened when the veterinarian describes Adonis's condition and the treatment he requires. Finally, Paul almost faints when he sees the bill for all this.

Because of this experience, we can understand why Paul vows to do everything he can to prevent a recurrence of the problem. Unfortunately, many chronic problems spring from some misalignment in that marvelously intricate balance we call feline health; and it often lies beyond our (imperfect) scientific comprehension. If we don't know exactly how a system works, let alone how it breaks down, we often conveniently and nebulously label the culprit "stress." We say that the individual who can't cope with stress will more likely succumb to the ubiquitous urinary tract or upper-respiratory viruses, or that kenneling or dietary or environmental changes can precipitate Sassy's diarrhea or aggravate Tommy's skin problems.

In addition to being a most vague and ill-defined term, the concept of stress yields another puzzling paradox. Cats not exposed to certain physical or behavioral stresses when young often mount extreme reactions when exposed to the identical circumstances later in life. In other words, some problems that plague adult cats result from the *lack* of certain stresses when young. Veterinarians expect to see roundworms in stool samples from kittens; when these same parasites show up repeatedly in an adult cat, we wonder what's wrong. Most believe that the changes created by the "stress" of early infestation enable the cat to avoid or eliminate similar parasitic assaults later in life. We noted how cats handled minimally (that is, minimally "stressed") as kittens reacted negatively and excessively to handling as adults.

On the other side of the coin, another raft of physical and behavioral problems result from the *presence* of stress. In our discussion of the various feline behavioral displays, we spoke of the stress created when one animal violates another's territory. Most people recognize the stress created when we change our cat's food, move with it to a new home, or introduce a new person or animal into the household.

Although the word *stress* crops up in the medical and behavioral sciences with a frequency that led one expert to call it "the diagnosis of choice for the diagnostically bereft," the fact remains that we don't know specifically what stress is. What stresses Adonis may not stress

Feline Science and Mythology: Dealing with Crises

Maggie or Copper. Moreover, what stresses high-pressure advertising executive Paul may not bother me or the low-key Harcourts.

"How can I hope to keep Adonis from getting stressed and coming down with a urinary tract infection if I don't know what stresses him?" wonders Paul.

He can't and neither can we. We can't get inside our cats' minds and bodies to determine what creates physiological or psychological incongruities for them. The most we can do is to venture some educated guesses based on our definitions of our normal cats. As a veterinarian I can tell Paul that kenneling stresses Adonis and that hiring a house- or cat-sitter might diminish this problem. Professional cat-sitters will visit the cat daily, interacting with it for an hour or two as well as attending to its basic needs. In Paul's case, this turns out to be a logical and acceptable solution for both him and his pet.

However, when another client adopts this same approach to eliminate feline skin problems that flare up in her absence, the result only complicates her problem. Not only does the cat resent the sitter's violation of its space, the owner really doesn't want strangers in her home either. In this case the solution actually produces more feline as well as owner stress (which the cat undoubtedly recognizes and responds to as well).

So, once again, we encounter the primary rule we must adhere to in order to deal successfully with chronic problems: *Do what best suits you and your cat*. Only by doing so can we eliminate the showers of negative emotions that ongoing but unpredictable problems can precipitate. Only by doing so can we provide a stable reference for our cats and a calming, stress-reducing influence in their lives. Although medically I may hold different opinions, I can't deny the fact that the Harcourts' choice to accept Copper's obesity and the chronic problems it precipitated did much to minimize the effect these problems had on their relationship. Because the Harcourts accepted the responsibility for these as the normal consequence of Copper's weight, they undertook most treatments with a minimal amount of emotion. Compare this to the owner of the fat cat who greets every bout of diarrhea or respiratory distress with, "Oh, it's all *my fault*. I never should have given him that lobster. How can I be so cruel to my poor baby? He deserves a better owner."

Facing the Unfaceable

Learning to recognize our cat's particular body-language expressions, their meaning, and our responses to them increases our awareness of this remarkable creature and our own remarkable ability to relate to it. While such awareness greatly expands our enjoyment of even the most mundane interactions with our cats, nowhere does it serve us so well as when catastrophe strikes. For most of us, the major crises come in the form of diseases such as feline leukemia, infectious peritonitis or cancer, or traumas such as car accidents or falls.

Unlike chronic conditions, these happen suddenly and quite often devastatingly. One minute the Harcourts are planning for Copper's next holiday photo, the next he lies comatose, fighting for his life. One minute Adonis curls up peacefully in Paul's niece's lap; the next, he slashes the four-year-old's cheek with his claws.

At times like these, owners understandably feel the situation slip completely from their control. When that first shock wave hits, we feel totally helpless and unable to do anything but go with the flow. Someone says, "Of course you'll take Copper to the XYZ Medical Center so that they can perform a heart-lung transplant," and the Harcourts nod dumbly. Paul's sister screams, "You *will* have that vicious beast destroyed immediately," and Paul mechanically obeys.

Unfortunately owners who abrogate their responsibilities at these critical times often suffer worse fates later. In the rush to do something *now*, and often strongly encouraged by others (including, and perhaps especially, some professionals), we often do the first thing that occurs or is presented to us. By the time the shock wave recedes sufficiently for the Harcourts to consider their relationship with Copper objectively, Copper is a hundred miles away hooked up to a heart/lung machine at the cost of several hundred dollars a day. By the time Paul calms down sufficiently to determine what happened, Adonis lies buried beneath the rosebushes in the backyard.

The death of a loved one taxes our strength to such a degree that we can't afford to complicate the situation with guilt and regret. Therefore, face the unfaceable now, while your cat is healthy; don't delay until a crisis forces you to stare death in the eyes.

"Oh no," shudders Paul. "I can't bear the thought of Adonis dying. It's just too terrible, too morbid. Besides, look how alive and healthy he is! I'll worry about death when the time comes." But so many of us fear death so deeply that the right time never comes, and we wind up floundering through what should have been one of the most meaningful experiences we can share with our pets—or even running away from it entirely.

There are some very practical as well as emotional reasons for confronting the issue of death beforehand. Many sophisticated and expensive advances have occurred in veterinary medicine in recent times. An evasive "Do everything you can, Doc" to duck personal responsibility when Sassy gets hit by a car may mount up to hundreds of dollars in a few hours time. Constant monitoring, X rays, ECGs, EEGs, blood tests, IVs, expensive soft-tissue surgery, exotic medications, and bone pinning or plating—much of which used to be available only in the most up-to-date human hospitals, now comprise routine options at many veterinary facilities.

Even if cost is no object, you need to recognize other limits, such as your personal life-style, that could pose problems. As a highly successful ad executive, Paul can easily say, "Spare no expense" when Adonis gets hit by a car or succumbs to malignant lymphoma. However, it's been my most sad and frustrating experience to discover that many owners facing critical situations often only consider either one of two possible outcomes: They expect the cat either to die or to return home perfectly healthy. They see the battle as between life and death; what *kind* of life will result is a distinction lost in the passion of the struggle.

Paul expects Adonis either to die from his cancer or injuries, or to

get better. He doesn't expect to have to take a half day off from work
every week to drive his partially paralyzed or sick cat to the university
for physical therapy or radiation and chemotherapy. He may not be
prepared to give his cat biweekly enemas because nerve damage sus-
tained in an accident makes it difficult for Adonis to defecate on his
own. If Paul doesn't consider his personal limitations objectively in
less turbulent times, he might find himself caught up in a situation
that robs him and Adonis of a quality relationship when they both need
it the most.

CONFRONTING ALTERNATIVES

In addition to the fairly cut-and-dried limits of finances and life-style,
we should also confront those created by our personal beliefs. For some
people, euthanasia never offers a viable alternative. This is a personal
choice reflecting personal beliefs, and others' judgment is irrelevant,
especially at times of crisis. If you reject euthanasia, make your position immediately clear to all involved in the cat's care and willingly
accept the responsibility for your choice. One of my blackest moments
in private practice occurred when an owner insisted that I hospitalize
his cat so that it could "die in peace." This owner's beliefs defined
euthanasia as inhumane, but his pocketbook convinced him that he
didn't want to treat the cat either. Still, he couldn't bear to "see that
poor creature suffer so." Unfortunately he expected his cat to pay a
very high price so that he could hang on to these three mutually exclusive and unworkable beliefs.

No, I'm not saying that all critically ill animals must be either treated
or euthanized, but I am saying that owners who refuse these options
should be willing to offer their cats emotional support and intimate
companionship during this time. What so frustrated me about this case
wasn't the owner's illogical position, but the fact that his own fears and
irresponsibility doomed that cat to die in a totally alien and frightening
environment without any of the compensatory benefits that environment had to offer. The cat would have been much better off at home,

Feline Science and Mythology: Dealing with Crises

in its box behind the wood stove. Although we can never know whether or how much fear intensifies pain in animals, it certainly does so in humans. The ability of many women today to give birth without the benefit of painkillers and anesthesia compared with thirty years ago directly relates to the decrease in fear that accompanied an increase in knowledge. Women who know what to expect and who feel comfortable in familiar and supportive surroundings relax and experience less discomfort than those who feel afraid and alienated.

Conversely, if you believe that euthanasia is the best route for you to take under the circumstances, make that position clear from the beginning too. Nothing is more heartbreaking for a practitioner than to give his or her all to save an animal's life only to discover that the owner doesn't share one's enthusiasm and commitment. Unfortunately the charged emotional climate that usually accompanies acute conditions makes this an all too common occurrence—and for two entirely different reasons.

Many times owners erroneously believe they really have no choice in the matter and leave the decision up to the veterinarian. Even worse, some people won't reveal this preference because they think the veterinarian will find them heartless. One client who put both herself and her terminally ill cat through eight agonizing weeks of treatment told me later she thought my professional oath made it impossible for me to accept euthanasia; and she hated the thought that I'd never speak to her again if she suggested it!

Although owners may maintain erroneous ideas regarding their veterinarian's response to euthanasia, some practitioners do deal poorly with this aspect of practice. Some won't even broach the subject, believing that owners will think them callous and uncaring. Others don't want to risk the outpouring of emotions the topic often releases. And still others see the death of any animal as a "loss" in the game of life, a symbol of failure on their part; a request for euthanasia is the equivalent to asking them to throw the game, and they want no part of it.

Fortunately few such practitioners exist, and they will become rarer as more and more veterinary curricula recognize the importance of exploring the emotional and intimate aspects of practice. To aid your veterinarian at these critical and emotional times, state your desires and limitations clearly. If you feel you lack the practitioner's support, objectively evaluate how this will affect you and your relationship with your cat. If you believe it will undermine what may be your last interactions with your pet or needlessly prolong any resolution of the problem as you define it, you owe it to yourself, your cat, and the practitioner to seek other care.

Other alternatives that often catch us unprepared at the time of crisis include the specific *kind* of care we want for our pets. Like human medicine, veterinary medicine no longer relies solely on antibiotics, anesthetics, and sophisticated technology. Although the traditional approach still comprises the mainstream of veterinary practice, alternate therapies such as acupuncture, herbology, chiropractic, homeopathy, and naturopathy have all attracted followers.

When a cat falls suddenly and critically ill, concerned owners immediately want to get the animal help as quickly as possible. For those who believe in the traditional medical approach, that poses little problem; many veterinary hospitals are, as one client noted, "just like human hospitals, except that the patients have more hair." However, suppose the Harcourts maintain some very strong beliefs about the unnaturalness or even the dangers inherent in all medications; they neither smoke nor drink and don't even keep aspirin in their home. How will they feel when Copper's treatment for congestive heart failure includes a regimen of five or six different drugs and all manner of technological paraphernalia? How will vegetarian Paul feel when his veterinarian tells him that Adonis *must* eat meat?

Two preventive measures can relieve owners of the mental anguish associated with such dilemmas. First, we can initiate and maintain a double standard: The Harcourts personally wouldn't be caught dead in a human hospital, but they don't extend that belief to the Maple Valley Animal Hospital. Although some purists may chide such as blatant

Feline Science and Mythology: Dealing with Crises

hypocrisy, many owners who embrace alternative forms of medical treatment for themselves can and do accept more traditional forms for their cats without experiencing any negative feelings.

Other owners who believe the alternative approach truly superior and want the same for their pets should arrange beforehand for it to be available in time of crisis. Sunday midnight is not the ideal time to track down a veterinarian skilled in herbology or homeopathy. Nor do you want to find yourself making your first forays into the application of alternative remedies under crisis conditions.

Although I deeply respect those skilled in the use of alternative therapies, I'll never forget my horror when a panicky owner rattled off the list of herbs and vitamins she'd given her cat during the preceding forty-eight hours in an attempt to abort its asthmatic attack "naturally." Although normally I would have treated such a problem without hesitation, I felt terribly inadequate in this case. My veterinary botanical knowledge barely extended beyond the ability to recognize and treat the effects of poisonous plants; what little I knew about medicinal herbs and their uses came from my interest in gardening. I wracked my brain trying to recall the active ingredients in the various herbs the client had given her cat, and I despaired over whether these were more or less potent in dried or fresh forms. Then I had to integrate the effect of the herbs with that of the massive doses of vitamins C and B the owner gave her cat to combat "stress." How much of these substances remained in the cat's system? And where? Finally, I had to decide whether my preferred choice of drug therapy would interact dangerously, or even work at all, in the presence of these other substances. All of this took time, during which the poor cat struggled to breathe, the sound of each breath cutting through me and the distraught owner like a knife. What should have been a relatively easy crisis for me *or* an alternative practitioner to resolve turned into a nightmare, the successful resolution of which owes as much to dumb luck and blind faith as any skill on my part.

We all want to believe that our daily care of our cats will ensure

THE BODY LANGUAGE AND EMOTION OF CATS

good health and reduce the frequency of problems and the subsequent need for treatment. But such crises do arise, and often at the most inconvenient times. If you ascribe to an alternate form of medical care for yourself and want the same for your cat, explore that option now. If you're going to do it, for your own and your cat's sake, do it right.

For further information regarding alternative approaches to veterinary care and the names of practitioners in your area send your request and a self-addressed, stamped envelope to:

> Dr. Carvel G. Tiekert, Corresponding Secretary
> American Veterinary
> Holistic Medical Association
> 2214 Old Emmorton Road
> Bel Air, Maryland   21014

DEMYSTIFYING THE MYSTIQUE

We began our discussion of the role body language and emotion play in crisis situations with the sad story of the famous fat cat, Copper, and his premature death as a result of his injudicious eating habits. As a newly graduated veterinarian pondering the sequence of events that culminated in that fateful day, my initial conclusions were adamant and simple: The Harcourts were wrong. Making food the keystone of their relationship with Copper set him and them on a course toward disaster from the instant his tongue scooped up that first dab of whipped cream. The basic foundation upon which they built their bond with Copper consisted of a mythology that flaunted every known fact of nutrition and animal care. Those were the brutal, irrefutable facts; nothing but flawed human mythology cost Copper his life.

As time went on, however, and Copper's memory flooded my mind at odd moments, my view of the relationship between Copper and the Harcourts altered. Initially my evaluation reeked of condemnation and

Feline Science and Mythology: Dealing with Crises

frustration. But soon I found myself fondly recalling Copper whenever another cat looked at me with his peculiar mischievous glint in its eye, that flash of gold across green that gave Copper his otherworldly air. Other times I'd hear owners crooning to their cats and remember the silly song Dottie Harcourt always sang to Copper—the song Copper stopped when he'd heard enough by carefully placing his paw gently across her lips. When tears sprang to my eyes as I watched nursing-home residents reach out to stroke their feline visitors from the animal shelter, I thought of how many hands of all ages had reached out to share Copper's special magic. Now I must concentrate to recall the specific ailments precipitated by Copper's extra weight, but the memory of his extraordinary personality and his owners' great love for him remains vividly clear.

This evolution of attitude in no way reflects my approval of the fat-cat mythology, any more than I believe that treating cats as human babies, dieties, or devils makes the most of the potential inherent in the human/feline bond. However, the afterglow of Copper's relationship with the Harcourts serves as a shining example of how much of *our* foolishness and mythology our cats will willingly endure for the pleasure of their bond with us.

Not long ago I received a brief note from Dottie Harcourt describing their new cat, Twinkle: "She's such a love, but nothing extraordinary like Copper, as far as others can tell. We only feed her cat food and hardly ever take her picture. And she'd never tolerate being dressed up the way Copper did. Just the other night Dick said she was the best cat. Oh, Copper was a great cat and the best one for us then, and he taught us a lot. But Twinkle is the best cat for us now."

I'm sure that Twinkle will live a long and happy life with the Harcourts, uninterrupted by the many medical problems that wove themselves through the Harcourts' bond with Copper. But it's funny, when I think of Copper now, I don't see an obese cat at all. Maybe I can now see what the Harcourts always saw: a slim, elegant angel of a cat, the epitome of the human/feline bond.

INDEX

INDEX

INDEX

INDEX

INDEX